Packaging design strategy

Other packaging books available from Pira

Advances in plastics packaging technology
Corrugated board packaging
Microwave packaging
Packaging and ecology
Packaging design and performance
Packaging materials

Pira packaging guide series

Packaging design strategy

Bill Stewart

Pira International

Reprinted 1995

ISBN 1 85802 064 6

First published by
Pira International
Randalls Road
Leatherhead
Surrey
KT22 7RU
UK

Tel: (+44) 01372 376161
Fax: (+44) 01372 377526

Typeset in the UK by Manton Typesetters,
5-7 Eastfield Road, Louth, Lincolnshire.
Printed in the UK by Hobbs the Printers,
Brunel Road, Totton, Hampshire.

Contents

1
Design concepts

As with religion, politics and education, everyone has an opinion to express about design. Encounters with design, whether it is in the form of packaging or any manufactured product, are an everyday experience. Consumers may be less than happy with a pack that dispenses badly, or they may wistfully admire the lines of a Porsche as they close the door of the family saloon. In both instances, they are ready to make some form of judgment, but what are the criteria that they use to decide what is good design and what is bad design? Clearly, in the instance of the pack that does not dispense, functionality is the major concern. If it does not perform satisfactorily, the design is at fault. A consumer's admiration for a Porsche is probably not based on personal experience; few people actually own one, and yet they are impressed by the overall appearance, and possibly influenced by its reputation for excellence.

When we assess design, we need to have some method of objective design measurement to augment our subjective view of what we like and dislike.

Decision making

The process of manufacturing even the simplest object entails conscious decision making by its designer. There are several decisions along the line for an object to be made one way rather than another. It follows that there must always be some design rationale, however idiosyncratic, behind all manufactured goods. This gives us a clue as to where to begin design assessment. We may or may not agree with the rationale, but at least it provides an insight into why the design takes the shape that it does, and gives us a measure of objectivity. When packaging design is commissioned, the design team must explain their rationale. Their thinking and decision making might otherwise be at fault, and lead them away from the desired objective.

Design, particularly packaging design, is normally a compromise, because it has many functions to perform. Successful designs perform these multiple tasks adequately (some probably better than others), while unsuccessful designs can be such a compromise that they do nothing well. While a consumer was admiring the Porsche, he or she might have been thinking how the children, dog and camping equipment would fare on a two-week touring holiday. Maybe an estate car would be the answer. Given unlimited resources, many people would end up with a variety of different vehicles to suit his or her various moods and needs. Most people are unlikely to indulge their tastes in this way. Similarly, it is unlikely that packaging will be originated that is dedicated to a particular task. On the other hand, if, when the design brief is written, the number of tasks defined for the pack are limited, there will be a greater chance of success. Some packaging functions — protection, producibility and cost — are features that have to form part of the design that will almost certainly have to be compromised upon. Wherever possible, additional features should be restricted to a minimum, and the real need for including them in the brief should be questioned.

Problem solving

There is a strong element of problem solving in design. Most design exercises start with a problem which the designer looks upon as a challenge. The problem should be encompassed within the design brief, but new problems often appear as the project progresses that were not foreseen at the outset. Design is therefore a process of problem solving, with the design being continually improved and moved forward. This is particularly true of structural design, where the ability to function well is paramount. Fortunately, such criteria are easily assessed. Simple tests for pourability, stability, and other physical properties inform us about how the pack performs. If some aspects are unsatisfactory, then modifications can be made, or the design can be rejected in favour of another, more suitable candidate. This may seem straightforward, but, in practice, there are many designed objects on the market which look good but do not meet the fundamental need to perform a task. For example, buying a teapot which dribbles down the spout is extremely irritating, but the purchaser will often persevere in its use because of the money invested. If someone buys a product with packaging that does not do what he or she believes it should do, that person tends not to buy it again, and the sales of the product decline. Before a pack gets into production, we should know whether it works. If there is any doubt that it will, we should consign it to the reject pile and move on.

Graphic design

While the physical attributes of packaging can be objectively assessed, it is far more difficult to judge its aesthetic appeal.

What is aesthetically pleasing to one individual may be quite unacceptable to another. If we make a distinction between functional structural design and styling, it is the styling that must be addressed. Graphic design is also very much part of the overall appearance of packaging, and so graphics are included.

All design is influenced by fashion. Occasionally, originality in design starts a fashion which is extensively copied and modified. Packaging is usually a follower rather than a leader of mainstream fashion, reflecting current trends, rather than establishing them. As fashion is so transient, there are dangers in following it and becoming locked into styling which rapidly becomes dated. Of greater significance in packaging is the ability of the designer to incorporate the desired product imagery. There are hardly any limits as to what can be achieved with this, but it should be borne in mind that, once a particular imagery has been established, it is often impossible to change it other than in an evolutionary way over time. Many of the major brands have evolved slowly over many years; it was not sensible to make radical changes which might alienate their target consumers. Their imagery should be updated using a step-by-step procedure which the consumer should hardly notice. It is only when the historical records are examined that the extent of this type of work can be appreciated. The evolution of the images of such well known brands as Camay, Coca-Cola and Shell make fascinating reading. This type of evolutionary design development may seem rather tame, lacking the excitement of original designs which break new ground. It is, in practice, one of the most demanding of design disciplines, and there are companies which specialize in this field. About 3M Mars bars are sold per day in the UK alone. When Mars decided to update this key product, they could not afford to change the design in any way which might alienate their core consumers. The brief, typical of its type, called for improved standout and a more up-to-date appearance, realized in such a way that the consumer would be unaware of any packaging change. The results were remarkable, in that the final design not only met the brief, but comfortably exceeded it, in one of the cleverest design studies of its kind. However, the Mars logo was altered in a subtle way. As this logo is probably the most valuable asset that the company owns, changes are not undertaken lightly. In this instance, Mr Mars had to approve and sign off the artwork in person (the company is still privately owned by the Mars family). This illustrates how packaging design reacts to fashion, usually by not adopting an overtly fashionable stance, but by following fashion and not being left behind.

Assessment of the success of such an evolutionary design is largely subjective, unless some form of market research is undertaken. It is even difficult to write a brief to express evolutionary change, as changes are required on the one hand, and no change should be apparent on the other. In the example of the Mars design, there were about 100 design experiments before the preferred candidate emerged. The design team had to exceed the brief to establish how far they could go before pulling back within the limits of the brief. Designers often challenge the brief in this way with a portion of their work, and it is entirely valid that they should do so. At the same time, the design was approached from the other end by incremental steps from the current pack. Close liaison with the marketing staff involved is essential to avoid carrying out work and therefore incurring costs in areas which are unacceptably far outside the brief.

Imagery

One of the most potent aspects of packaging design is the incorporation of imagery. This applies to almost all packaging, even of products such as ethical pharmaceuticals, where the end user may not have direct control over any purchasing decision. It is relatively easy for the designer to build in one dimension of imagery. For example, when the packaging of a jam is required to communicate qualities of the jam being homemade, a product of the countryside, traditional and English, this does not present undue problems, as the images are all compatible. The jar, if this formed part of the design exercise, might be squat, and the lid deep or covered by a gingham-printed paper disc. Graphically, for a countryside image, simple 'old-fashioned' typography might be appropriate if not considered too much of a cliché. Problems are likely to arise when mixed-imagery messages are part of the brief. In the jam example, if the marketing group also decides that the pack should have a 'continental' image, in an attempt to gain market share from imported French jams, then there is a conflict of imagery. A jam cannot be English and 'continental' with any degree of success, and, if this tactic is pursued, the design will suffer, and have a bland appearance which is neither one thing nor the other. The message should be clear: establish one type of imagery which is consistent throughout. Remember too that the more success the pack achieves, the more difficult it will be to change imagery midstream. Any future new-product developments that must be packed into the 'traditional' jam jar must be considered carefully. Suppose that the jam manufacturer wishes to market a fruit-based microwavable topping; we must consider whether 'microwavability' slogans and a traditional jar are compatible.

It is of great assistance to the design team and the marketing staff to build up a consumer profile. Designers can then understand what imagery is required, rather than having requirements imposed on them without explanation. The more detailed the profile is, the better. Even if much of the detail is subsequently disregarded, the design team will have had the opportunity of seeing for themselves what is likely to appeal to the target consumer. In a design study for a super premium motor oil, for example, the client was able to build up a picture of the consumer which captured the imagination of the designers. While the product itself was designed for high-performance cars, it was known that the owners of such vehicles rarely change their own oil, and, indeed, most of these cars do not require additional oil in-between regular services. The real market was the owners of older cars who were enthusiasts who cherished their cars, and were prepared to spend time and money on looking after them. Probably, most of these vehicles did not need a high-performance oil, but, for the true enthusiast interested in engine performance, this was a way in which they could cosset their engines. There was also the appealing concept of adding yet another performance ingredient. A bonus for the oil company was that older engines use more oil, and so revenues would be increased by repeat purchases. While this reasoning might appear calculating and cynical, there was no doubt about the quality of the product: it was all that it claimed to be. The important aspect was the aspirational nature of the product. The target consumer might not have been able to afford the latest BMW, but it was entirely within his financial means to provide his 'hot' family saloon with a high-performance lubricant. The target-consumer profile in this example was detailed: male, knowledgeable about cars, home owner, no drive, lager rather than real-ale drinker, holidays abroad on package tours, sports watcher on television but no participation, five-year old car, no trust in garages etc. With this profile, the designer can begin to imagine what is going to appeal to the consumer, and the imagery that the consumer is looking for in a product. A warning, however, is that the product has to perform to the standards of its imagery. It will not be successful for long if the product is all imagery and no substance. This particular design problem was satisfactorily solved by a functional, mechanical style of container that incorporated strengthening ribs similar to those on car engines: a fairly no-nonsense approach with little compromise. This study is discussed again in the book.

The ability to judge whether or not a particular design has the correct imagery is made much easier if a target-consumer profile can be compared with the design candidates proposed. Whether the marketing department has drawn up a correct profile is another matter.

PACKAGING PRINCIPLES

What packaging should do

At the most basic level, product packaging must contain the product, protect the product from degradation, and identify its contents (see Figure 1.1). This may seem elementary, but many packs fail in at least one of these areas.

By containment of the product, what we are really trying to achieve is efficient integrity of the package, so that the product does not leak, or fall out of or otherwise part company with its pack before it is supposed to. Remember too that the containment period may extend beyond purchase by the customer into actual product use. Once opened, a carton of milk, for example, should contain the product, and not, as some do, squirt it around the kitchen.

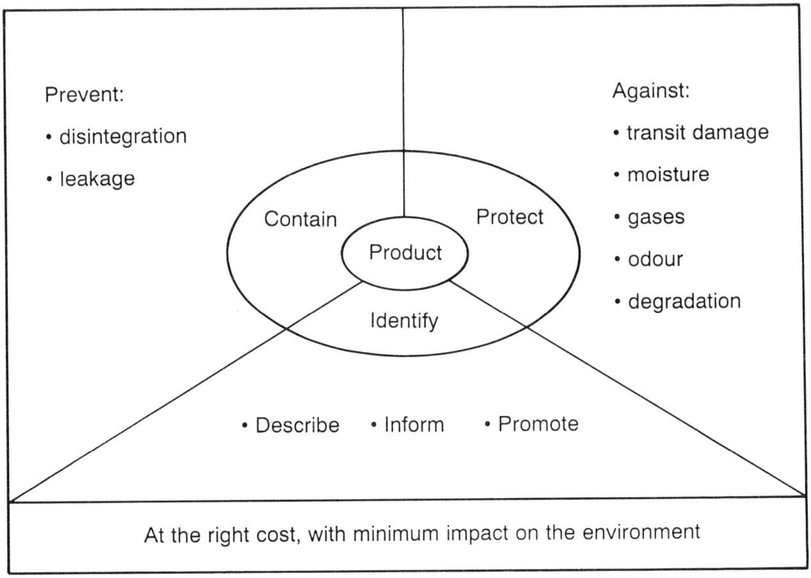

Figure 1.1 What packaging should do

The packaging has an essential role of protection to perform. It must protect the contents from degradation due to adverse conditions of, for example, moisture, gases, temperature, radiation, odour, infestation, and mechanical damage through handling and transit. The pack and the product must be compatible while this role is fulfilled.

The third basic element is the identification of the product. For some products, this simply consists of a description of the contents. In most instances, there is information about the product and its use, or a legally required text. The identification role may be extended to include promotion, where the pack not only identifies the contents, but actively promotes them.

These basic tasks must be carried out at the right cost with the minimum impact on the environment.

Containment

To explore what packaging can achieve, we need to expand the basic functions described above. If we take containment as a function, we can expand this to include dispensing, for example. The toothpaste dispenser not only contains the product, but offers a means of supplying it in a convenient and effective way. The pack has moved forward from the task of containment to becoming a vital part of the product. This can be taken further to cases in which the product itself cannot exist without the pack. Asthma sufferers and others whose medication includes the use of inhalers appreciate a packaging system which delivers the required dosage of the product in the most appropriate form for treatment. Here, the pack has become an integral part of the product, and one is dependent on the other. This extension of the containment function of packaging can apply to many types of product. Microwave cookery has evolved to include a wide range of convenience products which only function in partnership with their packaging. There is popcorn which is cooked inside a bag, the bag allowing heat and steam to operate as they would inside a conventional saucepan. Using advances in technology, some packs incorporate susceptors, which are areas of the pack that are coated with a metallic layer which responds to microwave energy by becoming hot. By the pack being 'printed' with metallic susceptors in carefully selected positions, it is possible to heat various areas of the product it contains selectively.

Such extensions of the basic functions of packaging are secondary functions; designers move on from what is essential to what is desirable, exploring what packaging can do. While new technologies play an important role in this process, for example by extending shelf life and providing consumer benefits, the more traditional roles of standout, imagery and branding cannot be ignored (see Figure 1.2).

The pack can be made to be an important contributor to product performance without an integral product/pack relationship being formed. Rigid flour packs, for example, allow

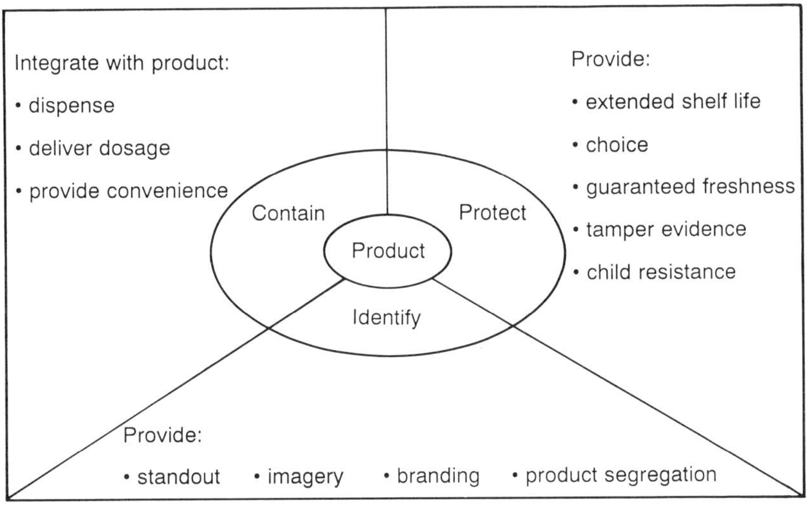

Figure 1.2 What packaging can do

easy storage and dispensing. It is all a matter of design coupled with accurate marketing strategy.

Protection

The protection role can also be extended. The 'tin' can was one of the first packaging techniques used to preserve foods. No doubt other methods of pickling in brine or sugar predate canning, but, with the advent of canned food, lifestyles were altered, setting a trend. The availability of food became nonseasonal, and choice increased. The possibility of preserving products, and protecting them against degradation, has led to many pack forms being developed. Supermarket shelves are full of products whose packaging, in the form of pouches, bags, bottles and many other composite containers, extends the shelf life of products, with obvious marketing advantages. There may also be marketing advantages in extending the role of packaging in the prevention of mechanical damage during transit or handling. While the basic requirement is product protection, there are instances in which, for example, the substitution of a plastic container for a glass one may increase product viability. The increased acceptability of polyethylene terephthalate (PET) bottles for carbonated drinks is partly due to the virtual unbreakability of the pack.

Identification

The product identification may simply consist of a label describing the contents. Packaging can build on this basic function by communicating in a far more immediate and interesting way. Content identification need not be just the result of graphics treatment. Pack shape can place the product in a particular category or even provide instant brand recognition. The classic shape and colour of the Perrier bottle can be recognized without the label having to be read. The consumer knows what the contents are after one glance. The original waisted Coca-Cola bottle is another example in which the shape identifies the product almost without recourse to graphics. Both bottle shapes have become unique brand properties of immense value to their parent companies, featuring in advertising and product promotion. The Toilet Duck bottle and Head & Shoulders container created, through shape alone, an association with particular market sectors. It was not long before competitors used similar styling, introducing 'me-too' products in bottles which approached that of the brand leader as closely as they felt was legally possible. The antidandruff-shampoo market became dominated by white bottles with extended necks. The consumer recognized both the colour and the shape as indicating that the shampoo was 'medicated' without the need for graphics. Similarly, a green bottle with a 'swan' neck became instantly recognizable as a lavatory cleaner. Reading the label was a secondary exercise, the potential purchaser having been directed to the desired category by colour and shape. The choice between brands became less clear, and decisions became more cost-oriented than was previously the case.

The primary use of graphics is to establish product identity. They tell the consumer what the contents of the pack are, either through text alone, or by a combination of text and graphics. The function of product identity is frequently overlooked, or at least not given sufficient emphasis. Consumers who may be inclined towards do-it-yourself activities, for example, need to know whether a drill is suitable for wood or metal, and exactly what size it is. They are looking for immediate satisfaction of their needs, and they are probably not inclined to sift through a mass of sales copy just to locate the vital information. If the details are clearly presented, perhaps by a pictogram backed up with text, their task becomes easier, and they become confident about making a purchase. The nature of the product may also dictate what information must be included on the pack. Instructions for use, the ingredients, a 'best-before' date, and the quantity, weight or volume are just some of the possible requirements. Mandatory textual requirements are likely to be specified by legislation, and other information may be desirable, such as environment-related details. This having been accommodated, attention can be turned to brand identity and the promotional aspects of graphic design.

What packaging should not do

The above discusses some of the things which packaging should do. There are also some things which it should not do.

Misinformation

It should not misinform. Packaging can misinform consumers by the impression of size that it creates, appearing to contain more product than is actually the case. There are legal implications here, and successful claims have been made in the UK under the law relating to misleading packaging. Cosmetic creams were a classic example, in which a hollow-walled plastic jar was considered to deceive the purchaser about the quantity of the product that it contained. Packs are still produced that are close to the borderline in this respect, but it is a mistake, perhaps even a legal one, to maximize the impression of size to the point of deception. Some products, such as cereals and powdered products, do settle after the pack is filled, occupying a smaller proportion of the pack. The consumer reacts, if this is excessive, by not making repeat purchases. Care must be taken in this area.

False promises

Packs should not promise more than they deliver. If a consumer sees a pack which suggests, through its design, that the product is in some way special, and then, after purchase, he or she finds that the product is in fact rather ordinary, the disappointment will be amplified by the elaborate packaging. The consumer feels deceived, and is not inclined to buy the product again. This is not just a product failure, but a pack failure too. It is difficult for a packaging-design team to tell the client organization that its product is poor or that the product concept is ill advised. It can hardly develop poor packaging to align with a poor product. What it has to do is to ensure that unjustified promises are resisted, pointing out that the imagery desired by the client might be unsuitable, and perhaps urging further market research before the project progresses too far. An example is that of a mass-market confectionery manufacturer who wished to enter the 'hand-made', fresh-cream-filled boxed-chocolates market. Because of production constraints, the product could not quite match the quality demanded by the market. The pack was designed with the fresh-chocolate market in mind, although care was taken that it should not be deliberately misleading. No doubt, however, it did overpromise, in the sense that the product did not live up to consumer expectations. It was not a success. It is, of course, almost impossible to say why the product failed, and whether it was because of the product, the pack, the marketing, the lack of advertising or the branding — all these must have played a part. Clearly, though, the overpromise of the packaging was one of the causes, and it is precisely this type of design that has to be avoided when one can exercise control.

Unnecessary packaging

Products should not be overpacked. This means that our aim must be to use the minimum amount of packaging that is appropriate to the task, and resist unnecessary packaging. For environmental reasons, only the minimum is acceptable, and, in practical terms, in any case, consumers react adversely to excessive packaging. The removal of layers of packaging is a task that consumers would rather do without, both because of the task involved and the problem of disposal. There is already considerable negative consumer reaction to packaging, and a sensitivity to excessive use of it. We need to exercise restraint when addressing packaging problems, and not be tempted to overkill, or, in this instance, overpacking.

Why pack at all?

Having considered the principles of containment, protection and identification, it may seem like heresy to consider doing without packaging. Posing the question 'do we need packaging?' should be part of every project, particularly in today's climate of environmental concern.

Fragility

Fragile items, once packed, enter the handling and distribution chain with their fragile nature disguised. Certainly there may be labels advising GLASS, FRAGILE, and THIS WAY UP, and there will probably be symbols too. None of this has much effect on how goods are handled. Handling is dependent on shape, configuration and weight. If the fragile article is in a box that is of a size and shape that is suitable for carrying at waist height, one can be certain that it will be dropped from this height during some stage of transit and handling. Anyone who has watched parcels being loaded and unloaded from the goods van of a train will have a fair idea of what can happen. If, however, the item is visible, and can clearly be seen to be fragile, even railway personnel may appreciate its delicacy and handle it with respect (although this cannot be relied upon).

This theory can work if the circumstances are right. For example, fragile car windscreens suffer higher rates of damage when they are packed in large and expensive corrugated cases than when they are visible. When paper covers the glass of the screen, and it has a foamed plastic surround, the curved shape is unmistakable. Car-body panels, furniture and 'white goods' (fridges, cookers etc.) are all examples for which the principle has been shown to work well, and, of course, to achieve a saving in packaging costs. These are all large items, however, and they are not really representative of the mainstream of consumer packaging.

Environmental factors

In supermarkets, there are currently few nonpacked items. Fresh fruit and vegetables and store-baked bread are usually the only goods sold on a self-service basis that are not packed. There are, however, may types of goods that need not be packed, and we become more aware of this if environmental issues are brought into the debate.

Consideration of these issues leads us to approach the question from a different standpoint, and think in a different way. If we want to reduce the level of packaging to help preserve resources and reduce energy usage, we have to consider using no packaging as a starting point. For example, flour, sugar, nuts, raisins, tea and coffee could be sold in bins, and the shopper could be asked to fill a simple paper bag with a scoop. This raises all sorts of problems of hygiene, product freshness and product branding. These are not unsurmountable. Even if bins are not acceptable, some form of dispensing unit could be used. Hard goods too, such as batteries, light bulbs, pens, pan scrubbers, and rubber gloves, could all be sold from a dispenser. Liquid detergents could be dispensed into reusable containers. While this is all possible, the effect on retail outlets would be tremendous, requiring a complete rethink of how stores are laid out and shelves replenished. It raises the issue of 'shrinkage', i.e. the loss of products through theft. Nevertheless, some of these measures are already on trial; the liquid-detergent example is being tried out in Germany because of the environmental benefits gained by reduced packaging.

It is easy for people working with packaging to assume, when working on a project, that some form of packaging is needed for a product, especially if they have spent their working lives designing packaging. It is now necessary to consider depackaging, which can be a cultural shock. It is right that we should do this, however, for both practical and environmental reasons. It also forces us to think in a different way, which is no bad thing.

ENVIRONMENTAL ISSUES

No packaging study should be undertaken without first considering the environmental issues. This is necessary not only on the grounds of pure environmental principles, but also to meet legislative requirements and deal with consumer awareness. Happily, the days are long gone when packaging the product in a green box was considered to constitute environmental soundness.

Packaging and the environment

Packaging has assumed a high profile within the environmental debate, partly because of the high level of visibility of packaging litter, and partly as a result of media attention to what is effectively an easy target that the public can relate to. After all, packaging is an everyday experience for most consumers, and they all have an opinion to express about it. In the domestic situation, packaging presents tangible environmental problems as people struggle to stuff their bins with discarded packs which resist every attempt to compress them, and prevent the closing of the lid. Bulging black refuse sacks outside front doors on bin day bear testament to the public's involvement with packaging, while consciences are salved with boxes of old newspapers and empty bottles in the garage that must be taken for recycling when time allows. It is hardly surprising that consumers have a largely negative attitude towards packaging, with it being regarded as a necessary evil, a regrettable part of the contemporary lifestyle, and an unnecessary cost.

Despite the fact that many people are too young to remember the days when biscuits were sold loose, the public have an underlying desire to move back in this direction. Those consumers who can remember those days realize the folly of this rather romanticized view of the past. In truth, the biscuits were quite often stale, soft, and sold in conditions which would require the employment of an army of health inspectors. Lifestyles then were very different from those now. Selective turning back of the clock will not work, and a host of modern values must be accepted. In many parts of Europe, biscuits, and many other categories of product, are still sold in the traditional way, but this is matched by a compatible lifestyle in which, for example, women shop daily. This is no longer an acceptable or even possible scenario in the UK.

Communication with the consumer

Product manufacturers that are serious about wishing to limit damage to the planet caused by packaging must consider the real environmental issues underpinning the design and selection of packaging, and how any consequent benefits can be brought to the attention of the consumer. After all, if environmental factors influence a purchasing decision, it makes sense to be seen to be 'green'. Rows of packs claiming environmental benefits now confront the consumer, but few of these assist the customer to make an environmental decision. At least, however, manufacturers have initiated some progress towards an environmental goal, which, in the light of today's ready acceptance of compromise, represents a degree of achievement.

Detergent manufacturers have recognized the greater consumer awareness of environmental issues, and the need to explain aspects of these issues to consumers. Television advertising costs serious amounts of money, and the extent of explanatory advertising is a measure of just how seriously the manufacturers of concentrated detergents and top-up packs take environmental matters, and how important they must be to sales. This move towards packaging which at least accords the environment a higher priority is good news for everyone. It begins to take the burden of decision making away from the consumer, who can relax a little, knowing that manufacturers are shouldering the environmental burden, leaving them free to select products using traditional criteria: price, quality and individual preference. In the same way as consumers expect biscuits to be fresh and packed in hygienic conditions, and pharmaceutical products to be safe, they will begin to expect the packaging of goods to minimize any detrimental effect on the environment. This will defuse some aspects of consumer concern about the supply of goods, and shift the focus of the environmental lobby towards disposal, where the active participation of the consumer will be required.

Waste disposal

Packaging designers, marketing staff and manufacturers must be aware of and respond to the challenge of producing environmentally responsible packaging, and react to the changes in consumer focus by considering disposal methods. In fact, it is necessary to cast a critical environmental eye over the whole production process, from beginning to end.

All human activity generates waste. In the UK, it is estimated that around 4% of solid waste by weight comes from domestic sources. This is an estimate extrapolated from local-authority measurements, although, perhaps because waste disposal has been less important in the past than it is now, many local authorities still do not produce figures for weights of waste.

Domestic waste, i.e. the contents of domestic waste bins, are all too familiar, typically containing potato peelings, vegetable trimmings, discarded packaging, and an increasing amount of junk mail. The packaging content has been estimated to be around 1% by weight of all solid waste, as shown in Figure 1.3. The percentage by volume is very different, because the bulk of domestic waste is formed by packaging. The composition of domestic waste reflects lifestyles, and, as lifestyles change, so does the composition and amount of domestic rubbish. This is true for all countries: different styles of living create differing patterns of waste. Varying levels of environmental awareness also affect the makeup of domestic rubbish. Refuse from a bin in a German home differs from that of a typical UK household, reflecting aspects of lifestyle and environmental differences. It would not be uncommon in Germany for tea bags to be separated into their paper and used-tealeaf components, and disposed of in different bins. Consumers in the UK, the bastion of tea drinkers, would never dream of doing this. This example illustrates the difference in environmental awareness between two countries that are similar in many other ways.

As shown in Figure 1.3, agriculture creates the most waste, followed by industry. The fact that accurate waste statistics are difficult to obtain reflects a lack of commitment to environmental concerns, but this is rapidly changing, and impending

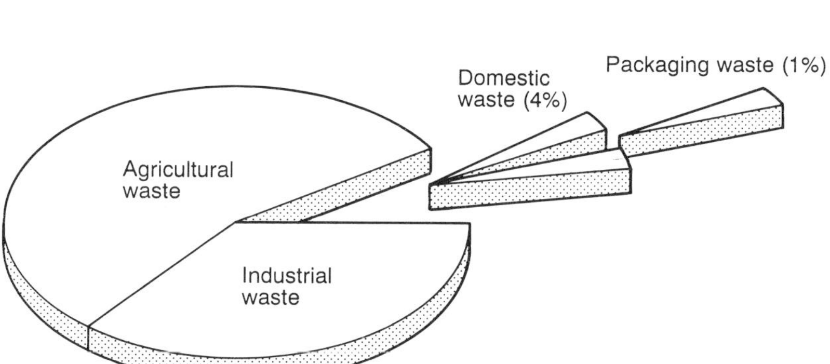

Figure 1.3 Packaging waste as a percentage of solid waste, by weight

European Commission Directives will call for the monitoring of waste quantities and processing.

Materials

The environmental question most often asked of packaging designers is that of which is the most environmentally friendly material to use. There is no simple answer, except that one must seek the material that is the least environmentally antagonistic; all materials impact on the environment.

Environmental benefits

Packaging invests the energy used in its creation in protecting the product against damage. Product spoilage would result in the waste of the energy used in the production of the product, and further expenditure of energy in its replacement.

Packaging helps to provide product choice. The preservation of products which would otherwise swiftly degrade enriches the lives of consumers. The packaging of food, for example, allows products from around the world to be sold nonseasonally.

Packaging provides convenience when consumers' lifestyles place heavy demands on their time, allowing them to buy provisions once a week, once a month, or at even longer intervals. Ease of storage, and minimized preparation and cooking times for foods fit in with modern living patterns.

Packaging helps to maintain health and hygiene by preventing the contamination of products by bacteria and toxins.

The environmental factors that must be considered during packaging-design projects are as follows:

- Does the product need packaging at all? All packaging can be a burden on the environment.
- Are the product and manufacturing processes environmentally sound. There is little point in using and producing environmentally acceptable packaging if the product and processes are environmentally unsound.

- Do the benefits of the environmental packaging performance outweigh the disadvantages?
- Does the packaging meet all the conventional performance criteria in addition to the environmental criteria?
- Have the recycling problems been minimized by the avoidance of the use of mixed materials? Have the materials been identified, and pigmented materials avoided?

The profile of the ongoing environmental debate will change in line with other events that achieve media dominance, and it is a debate that is not going to disappear. In the packaging business, the originators of packaging have a primary responsibility to use their knowledge to the best advantage, ensuring that only the minimum amount of packaging is used, in conformance with the principles of packaging design. More will be learned as time goes on. There will be a greater understanding of environmental issues, and more accurate information will be available on which judgments can be based. We must make sure that we have the flexibility to take this new information on board and use it, on behalf of designers, marketing staff and consumers, to the benefit of the planet as a whole.

INTEGRATED PACKAGING

Packaging design cannot be considered in isolation; it has to be integrated with other disciplines. There is little point in designing a pack which is expensive to manufacture, difficult to fill, and does not meet merchandising requirements. The designer, unlike the artist, is not free to exercise imagination without constraint, and cannot simply retire to the studio and produce a design masterpiece. Although the client may be represented by one or two individuals, often from the marketing department, there are, behind the scenes, further disciplines which need to be satisfied. They too will have opinions about packaging, and expertise within their own field which can only enhance and enrich the packaging study. It is often the case that such views create contradictions. The packaging buyer may want the flexibility of multiple sourcing, while the production engineer may wish to use a packaging system that is available from only one source. These contradictions can be resolved, and the right level of compromise reached. Apart from the necessity to identify the requirements of each discipline involved, there is a very real benefit to be obtained by involving all the relevant departments in a packaging study. Any programme for change within a company will be more successful if the support and positive involvement of all departments can be achieved (see Figure 1.4).

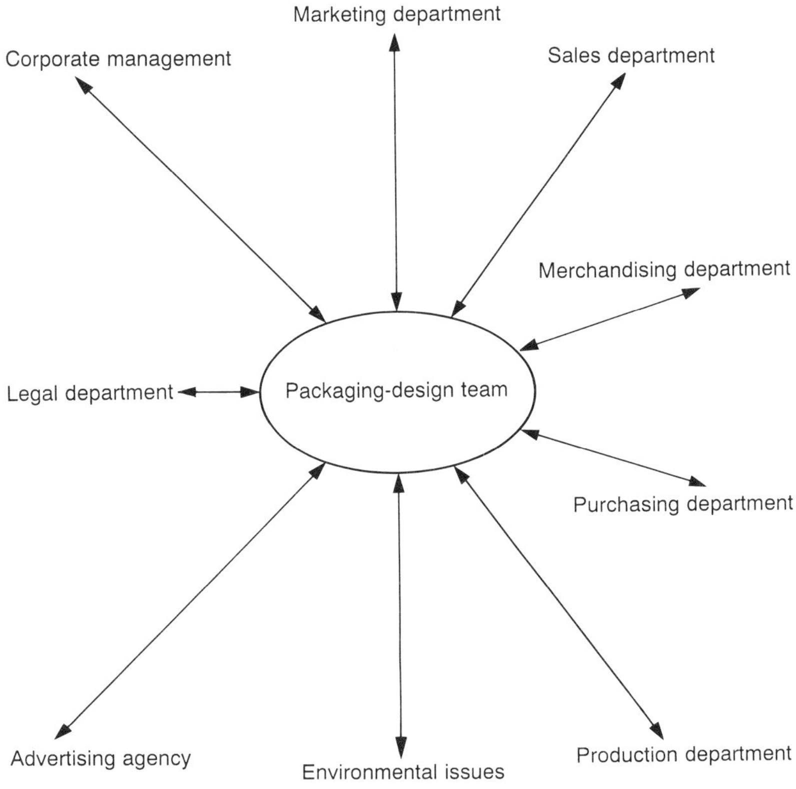

Figure 1.4 Integrated packaging

Purchasing departments

Purchasing departments, and specifically the packaging buyers, have the task of ensuring that the right packaging materials are delivered at the right time, to the correct specification and at the right price. To achieve this, they will build a rapport with favoured suppliers, and gain an in-depth knowledge of the materials and manufacturing processes relevant to their speciality. A design study which ignores this expertise, or presents a *fait accompli*, misses out on the benefits of a valuable resource. Whenever possible, packaging designs should allow the buyer freedom to obtain quotes from a number of suppliers. Some companies insist, as a matter of course, on being able to source packaging from two or three suppliers. It is, in any case, good policy to use competitive tendering for cost reasons, and to ensure continuity of supply should a supplier fail to deliver because of financial, physical or other disasters. The design team which provides a solution which effectively confines the buyer to single sourcing may risk the client's overall business interests. There are few occasions when this is justified. Even when a pack design is produced on licence, there should still be scope for multiple sourcing. The most common grey area is that of tooling. If tools have to be created, say for an injection-moulded pack, the tool costs are likely to be substantial, and the provision of duplicate tooling at additional suppliers may be cost-prohibitive. While the contract to produce packaging is open to competition, the selected supplier, in the instance of one set of tooling being desired, is in a favourable position. This is a commercial risk, and the buyer will have to be very sure that the right company has been chosen. To reduce tooling costs, there is often an arrangement whereby the supplier bears part of the tool-origination costs. Although this is attractive initially, the system starts to fall apart as soon as there are problems of price, quality or delivery. It is better for the client company to purchase tooling outright, and to have the flexibility and bargaining power to move tooling between suppliers. In practice, however, this option is often not open, as tools are usually constructed to function on one particular type of machine, and may not be interchangeable between packaging suppliers.

In the early stages of a design study, these considerations should not hinder the creativity of the design team, but early discussions with the purchasing department help to avoid later pitfalls. They also, importantly, start to build up an open relationship between the design team and the purchasing department, which will be to the overall benefit of the study. There are also instances in which the pack form being promoted by the designers may not be within the packaging buyers' current portfolio, because it uses materials which have never been purchased by the company before. Here, the buyer will look for support from the designers in identifying potential sources of supply.

Suppliers

In any design study, there will come a point when the designers need to involve suppliers, usually to obtain outline costs or detailed production information. Again, the client's purchasing department should be involved in establishing which suppliers to contact and which individual within the supplier company is likely to give the best service. Equally importantly, the buyer can advise on which companies should not be contacted, perhaps because of previous bad experiences or company policy.

With some projects, particularly those where new packaging is being developed or new products are involved, there is a requirement for absolute confidentiality. The world of packaging is a small one, and word of new enquiries soon circulates, perhaps alerting competing client companies to the client's future intentions. There is more than one example of new product intelligence being gathered via packaging suppliers. The nature and source of the project may be masked by using a designer or agency to make the initial supplier contact. The disadvantage of working in this way is that suppliers are unlikely to respond with much enthusiasm to design agencies, and certainly not in the way that they do to a major prospective customer. This is reflected in their attitudes and their quotations. Prices obtained in this way are still relevant. A design agency is unlikely to be quoted the same price for a pack as Procter & Gamble, for example, but at least the design solutions can be ranked in order of price.

Production department

The production manager will be less than pleased to be presented with a pack which is going to be a nightmare to fill. Designers should become familiar with the processes and machinery available within the production department before design work is started. This need not necessarily dictate the pack form, but it should at least be a consideration. There may be valid reasons for considering the installation of new equipment to handle an innovative new pack design. Whether existing or new machinery is involved, the production department needs to become an integral part of the project and to be consulted at an early stage. Certainly, in larger organizations, the production department will be a cost centre, having budgets and targets to meet. There can understandably be resistance to design concepts which increase costs or are a mismatch with machinery. In contrast to the commonly held belief, production departments are often keen to make changes. They have identified the shortcomings of their current systems, and may well be receptive to concepts which improve their methods. The important thing is to involve production departments. Their views and opinions are valuable. Only by obtaining the department's full support can a project progress smoothly. Too often, production departments are considered to be a necessary, if inconvenient, adjunct to the selling of product. This attitude can only damage the overall performance of a company. Particularly in the introduction of new packaging or products or both, it is vital to have the approval and whole-hearted support of the production staff. Packaging design needs to address machinery issues, the implications for capital costs, and the effect on line speeds. The handling of packaging through the lines, filling, capping, storage and transport are all affected by package design, and the impacts on them have to be assessed.

Advertising agencies

When major packaging exercises are undertaken, it is surprising how infrequently the advertising and design agencies coordinate their activities. This is a pity, because advertising agencies often have a slightly different perspective on the role of packaging which can form a valuable input to the study. Many advertising campaigns feature the pack, and it makes sense for designers to listen to the opinions of the advertising agency. Perhaps it is because of the need for advertising to be imaginative, but advertising agencies are particularly good at building up a picture of the product positioning and target audience which is readily understood by designers. Both disciplines share a common language.

Legal department

Some companies and certainly the major multinationals, have a legal department whose activities should be integrated with the packaging design process. All copy should be checked and signed off by the legal department before it is passed to the design studio. Artwork, which is often the final stage before production, also needs legal clearance. This has to be done as a check that the copy is accurate, and to make sure that the final copy size and positioning is correct. Any photographs or illustrations similarly need to be approved to ensure that they are not misleading or likely to cause legal problems in the country of sale. If the project has involved 3D design, there may be questions of misleading impressions of size which will require legal guidance. During the course of a project, there may also be other issues related to patents and copyright which may affect 3D and graphics designs. It is normal for design companies to assign copyright or patent rights to the client company. This may result in a small substage in a project in which drawings are prepared for submission to the appropriate agencies, in the case of multiple-country applications, involving a considerable amount of work.

Environmental monitoring

All packaging design must be integrated with some form of environmental monitoring. Some companies already have an office that is charged with this responsibility that may also be involved with packaging matters, and others will rely on guidance from the design studio or employ an external specialist. Whatever method is used, it is essential that the impact of packaging, whether existing or new, be considered in this way.

Marketing, sales and merchandising departments

The prime contact in any packaging study is with the client's marketing group. They are normally the driving force behind packaging studies. Some groups include sales and merchandising staff as an integral part of their operation. Other organizations separate the functions. Merchandising is closely related to packaging activities, and, where there is expertise within a company, it should be brought to bear on all packaging work, and this is essential if any new pack forms are considered. The sales group also have a part to play. They are at the sharp end of the business, and their viewpoint may differ in detail from that of the marketing department. It is always worthwhile obtaining their opinions on any new or revised packaging designs. They have a knowledge of the customer and the end user. Their customers may be major multiples, and they may have to sell into this type of outlet. A knowledge of what particular buyers are seeking in terms of pack design could create an important competitive edge.

The benefits of opening out design work to a wider audience are significant for the success of the final design, and the ongoing relationships within a company. Integrate the process of packaging design, and everybody benefits.

PACKAGING-DESIGN STRATEGY

The Greek shepherd roaming the mountains with his goatskin bag of wine had no idea that, thousands of years later, branches of Tesco in the UK would use the same principle to sell bag-in-box wines from around the world. The originator of the goat-skin-bag design obtained nothing in the way of a patent, and history does not accord him or her celebrity-designer status. Then, as now, packaging had to perform the crucial task of protecting the product, but the goatskin also, cleverly, dispensed the product, giving it a marketing advantage over other pack forms. Whether much in the way of market research was conducted is lost in the mists of time. It would be good to think that someone sat down and analysed the need for the wine to be portable and fresh, and access to be easy and breakage impossible. Of course, the designer had no marketing, production, purchasing, legal and advertising departments to contend with.

Today, we recognize the importance of packaging, not only as a means of protecting the product, but also as a means of selling it. We cannot afford to sit around and wait for inspiration to produce an innovative pack form, and indeed innovation on its own is not enough. Designers have to see a gap in the market, and design products and packaging which will fill it.

While the ability to design is a product of creative thinking, the ability to achieve success is dependent on analytical thought. Creative genius is not sufficient — it has to be strategically directed towards a final goal. The design has to meet all the criteria imposed by the many disciplines involved, not only satisfying the marketing department, but also the purchasing, legal, production, advertising and merchandising departments. The designer of the goatskin wine pack would have experienced a rough ride in today's climate.

Where success is measured in terms of sales and profit, good ideas are not enough. The designs have to perform. For the achievement of success, inspiration on its own is insufficient; creativity has to be harnessed to an analytical approach. This is strategic packaging design.

By the use of a step-by-step strategic approach, creativity can be directed towards practical design solutions that work, judging success by sales and not design awards. The goatskin wine bag could have won a few awards, but it certainly could not have established a brand.

Project management

To make a packaging design project work, it is critical that the task in hand be identified. What exactly is it necessary to achieve? Firm objectives need to be set, forming the vestigial skeleton of the brief. The flesh can be put on the bones later, as described later in the book. It is both convenient and practical to break the project into stages, and create a planned order of the events that must happen for the project's conclusion to be reached. This needs to be reviewed to establish what departments and which individuals are going to be involved, and which of them will have key roles. Communications methods should be considered even at this early stage. It may be necessary to hold regular weekly meetings throughout the project, or some other arrangement may be more satisfactory. The appropriate communication system must be put in place to ensure that all those involved participate fully, and do not take a back seat and then criticise the results without offering solutions. The project will be a team effort, and all those involved must be prepared to air their views openly. One major international company has adopted the maxim 'don't present problems, provide solutions'. This is entirely appropriate for packaging projects, and all the team members should be made aware of this philosophy from the outset. Problems will undoubtedly arise, but the team must not dwell on them, just overcome them. No project can be planned without establishing costs and timing, even if only in outline at the first stage.

This systematic approach to a creative process is the essence of packaging design strategy, and each aspect of this approach is covered in greater detail in subsequent chapters.

2

Design sources and budgeting

In the 1960s, there were no specialist packaging designers that were based in the UK. There were generalists. If someone wanted to commission packaging design, he or she might approach suppliers, industrial designers or packaging engineers. By the mid1980s, however, all this had changed. There was a new emphasis on design at every level and in all areas of life, from designer clothes to designer cars. It seemed as if the consumer society had embraced design in all its manifestations, including packaging, on the face of it an unlikely candidate for centre-stage stardom. Some design consultancies had figureheads who achieved 'personality' status in the media, becoming the gurus of good taste in design. Consultancies, basking in the reflected glory that accompanied such euphoria, expanded quickly, and increased their fees to staggering proportions. Designers were all going to be rich!

When the bubble burst, and it was realized that people had been paid too much for too little, many consultancies and their respective gurus quietly disappeared. Manufacturing industry also felt the cold draught of recession, going out of business, selling out to larger groups, or soldiering on, hoping to weather the storm. From being a high-profile activity, design slipped back into its supporting role, where it is likely to remain until the cycle once more brings it to the fore.

The froth has been blown off the top of packaging design, and only pragmatic considerations remain. One of the lessons to be learned from this is that, when a design source is selected, a strategic policy must be adopted. An agency must be chosen that provides the right pack at the right price at the right time. That does not mean that the cheapest source or a source that is renowned for winning design awards should be chosen, but a source that is the best suited to fulfilling the brief, and delivering the goods on time and at a fair price.

A packaging project should be described in a brief, and in a later chapter there are guidelines for making sure that the brief is adequate. Where can the marketing manager, brand manager or other individual responsible for the project then go to achieve the best value?

There may well be constraints which limit the choice. The budget available is often the first major constraint. There can also be design constraints, for example a logo designed by the chairman's wife. The project needs to be aligned with the correct design source. The first step is therefore to select the type of design source that is most suitable. Then specific companies can be evaluated.

IN-HOUSE DESIGN

Larger organizations often employ their own internal design team. The level of design work undertaken by such in-house sections varies from largely technical or packaging-engineering activities to graphic design. It is perhaps more usual for a manufacturing company to have a packaging-engineering section, where the technical, rather than design, aspects are considered. Packaging engineers possess differing degrees of design capability, but their educational backgrounds are probably in technology rather than design. Nevertheless, engineers and technicians are not precluded from being good designers, and engineers are surprisingly often adept at design. This is particularly true of structural engineers, and an engineering-based structural-design solution can be far more elegant and less contrived than one developed by a designer who is preoccupied with styling rather than function. It is, however, much rarer for packaging engineers to possess graphic-design skills.

The members of an in-house design team operating at any level of design competence have an advantage in terms of their immediate understanding of the product and production techniques. They are also likely to be in contact with suppliers, and to have an established rapport with them in relation to a variety of issues.

When marketing staff judge that a project requires creative flair, in-house design teams are seldom consulted. It may be that the credibility of in-house teams as creative designers has not been established to the same extent as their technical skills but, rightly or wrongly, they are rarely awarded the leading role in any design exercise. They may become so close to the product and its production that it becomes more difficult for them to take an independent view in the same way as an outside agency can. Rarely, however, are in-house design teams briefed in the same way that an outside agency is. Often, too, they are not given the time or resources to conduct research or evaluate the market. In other words, they are seldom provided with the information and resources that would enable them to compete effectively with outside agencies. They are also not able to remain available until a design project arises. Their time is spent on many other aspects of the company's activities.

In-house design teams are thus not well placed to take on packaging-design projects, which is a pity, as there can often be much latent talent in the team which remains unexpressed. In-house knowledge and outside design flair, however, is a powerful combination. Not all such arrangements work satisfactorily, because success depends on the personalities involved. Where there is compatability, the result will be creative and yet practical design work.

COMPETITIVE TENDERING

Many companies, particularly when economic conditions mean that design consultancies are desperate for work, employ three or more consultancies during the first stage of a project, with a promise of work on further stages for the agency that performs best. The marketing budget for the first stage is usually, in this situation, divided between the participating design studios. This means that each receives a low payment in relation to the amount of work involved. For example, a job that would justify a fee of £15 000 in terms of the hours needed to solve the design problem is paid for by a fixed fee of £5000 per agency. Most design consultancies will take the view that it is worth competing, either to gain a new client or to gain further design stages on the project. In attempting to win the competition, each consultancy inevitably exceeds the set budget, doing work that it knows it cannot bill. However, it hopes to recoup the costs by increments during further stages. It would be possible for a design agency to spend ten times the amount of billable time allowed for by the client's budget on the project.

On the face of it, the client is getting a good deal, acquiring three versions of a first-stage study for the price of one. Also, the keen edge of competition ensures that each competitor will have conducted a fairly intensive study. However, is it really such a good deal?

It is difficult to judge design. If each agency recommends two or three routes for further development, there will be about nine different designs to be considered. If the agencies have done their work properly, each route will have been carefully thought through, and the rationale behind each route will have been detailed. If there is clearly a winning design, then there will be no problem. It is far more likely, however, that one design will be strong in one area, for example standout, while another will have other desirable characteristics. The dilemma then is how to take the best from each design in some form of crossfertilization. This can be tricky enough when only one consultancy is involved. When such crossfertilization is needed between designs from different agencies, problems will inevitably arise. Taking a feature from one company's design and handing it to another for development is generally unsatisfactory. Although each consultancy working on the project will not know what the others have proposed, it is particularly frustrating for designers to see elements of their work appearing under another company's logo. There have been instances of legal action as a result of this practice. While clients have a right to use whatever work they pay for, instructing a design group to develop work which they have not originated, and therefore may not fully believe in, leads to lack of commitment. It is far more satisfactory to select a design group before commissioning work, or, when there is a competitive first stage, to use only the work of the preferred agency. Interagency crossfertilization should be avoided. Apart from the ethical issues, the rationale behind each design element becomes confused in this situation, and this tends to weaken rather than strengthen design quality.

LONG-TERM RELATIONSHIPS

Many larger manufacturing organizations have established long-term relationships with selected design consultancies, and they often maintain a design register which is subject to performance review. In some cases, these relationships are a result of the rationalization of departments within major companies, leading to packaging activities being contracted out. Other companies have packaging sections that are oriented towards packaging engineering rather than design.

An appointment as an approved design house retained by a large organization is a sought-after accolade. It offers an opportunity for the consultancy to be involved semipermanently, and it guarantees a level of income. This is an argument for the client to negotiate special rates. There are operational benefits too, in that the studio becomes familiar with the client's products, its production methods, and its organizational structure.

When they work, such arrangements are to the mutual benefit of both parties, and the design studio effectively becomes an extension of the manufacturing company's own staffing structure. With the trend being towards fewer but larger manufacturing companies, many consultancies see this process as being their best way forward. There is an argument also that it is becoming increasingly difficult for consultancies to avoid conflicts of interest when so few manufacturers dominate the market. It may be, for example, a problem to work both for Procter & Gamble and Unilever, as they are in direct competition in many fields. If a manufacturer demands an agency's exclusive loyalty, it is, in effect, setting up an exclusive arrangement, however informal the agreement is. Often, a secrecy or confidentiality clause may be the only loyalty agreement between the two parties. However, with fewer manufacturers holding larger shares of markets, the problems faced by smaller organizations in dealing with them must be recognized, and this may apply to other small suppliers also.

Rather than being too dogmatic, most multinational manufacturers are coming to terms with using suppliers who also work for their competitors. Provided that the competition is not direct (for example the design of shampoo labels for both companies), then there is little problem. Deciding where to draw the line is not easy, however. The management of studios will often persuade themselves that they can safely work on two projects simultaneously, erring perhaps on the side of optimism. For example, they may persuade themselves that antidandruff shampoo is sold to an entirely different market from that for protein-enriched beauty shampoo. Shampoo manufacturers may not agree, and they may interpret this type of comparison entirely differently. Long-term agreements avoid this situation.

There is a danger, however, that a design house working exclusively for one company or on one brand for a large organization may become stale. It can become very comfortable for a consultancy to be part of a larger group. Familiarity with products and staff does have a downside. The design-house thinking is soon aligned with that of the company, and design judgment may not be as challenging as it should be, and creativity may thus be stifled. Designers also become bored when working for a long time on the same project, particularly if the work consists of small changes that are largely variations on a theme. Bored designers produce boring designs that lack the sparkle which the designers were originally employed to create. A degree of competition, or switching consultancies on the design roster from one brand to another, helps to maintain a fresh approach. In practice, this seldom happens. Usually, in large multinationals, there are brand managers who are responsible for the design of a limited range of products. Each brand manager appoints his/her favoured consultancy, and tends to stick to the selection. Often, brand managers do not engage in crossbrand discussions, and do not offer 'their' consultancy for the use of the others in the group. The introduction of competition, by alerting consultancies to the fact that their role is subject to periodic review, helps to keep designers on their toes, and alerts account managers that costs must be kept under control.

SELECTING THE RIGHT CONSULTANCY

It is vital to match the project in hand to the capabilities of the selected consultancy. Consultancies inevitably develop a house style that not only reflects the particular talents of the designers, but also the judgment of the creative director and the management team. It is this that largely dictates the type of work which the consultancy specializes in and, not surprisingly, the work that they are good at. It becomes difficult for consultancies to move out of the niche that they have created for themselves, no matter how hard they try. From a potential client's point of view, this is no bad thing, because at least it eases the task of consultancy selection. Specializations are often very narrow. It is usually by product type (for example wines and spirits). It may also be by type of project undertaken (for example design updates for brand leaders).

Most consultancies employ a sales and marketing team to promote their range of services. Mailshots and public relations activities often overstate the capabilities of the agency, claiming that it specializes in a wider range of work than is actually the case. A more accurate assessment can be made by examining the agency's portfolio with a critical eye. This will reveal the successful areas of operation, and provide a clearer insight into its true line of specialization. Consultancies are usually only too pleased to arrange a credentials presentation to any prospective client, and they will not only show examples of work, but also explain their approach to projects. This is equally important, as a disciplined methodology is more likely to create positive results than inspired but misdirected creative effort, which the client will have to pay for. If at all possible, initial presentations should be given at the design studio, where the client will have an opportunity to meet the key staff who would work on the project, rather than just the sales team. The client's staff may have to work alongside creative staff for a considerable time, and, if personalities do not gel at the initial meeting, the chances are that difficulties will arise at a later, more critical stage.

When suitable consultancies are being sought, it is important to have fairly well established ideas about the project in hand. Initially, there may not be a full brief, but there must at least be sufficient detail to aid the selection of the right studio. The nature of the project, whether it is structural, technical or graphic, must be known. Most design agencies claim to offer all of these. The agency's recent work must be closely examined, and the split between these disciplines assessed. If the potential project has structural or technical elements, it must be established that there are sufficient talents within the agency to cover the practical issues, and that none of the work will be subcontracted. Skills are required for the development of a brand leader that are different from those required for the introduction of a new brand or product. For example, studios have become adept at updating brand images so that standout is improved while the brand is still instantly recognizable to consumers as their favourite brand. UK-based consultancies are usually well versed in the requirements of the UK market. If, however, the product is going to be sold in other countries, it is essential to determine the studio's familiarity with local market conditions. All packaging studies contain an environmental element. The environmental track record, and not the claims made, must be considered.

The above factors, which are summarized in Table 2.1, comprise the basic level of information that is needed when a design consultancy is approached with a view to commissioning a project.

Table 2.1 Information needed when a design consultancy is approached

Project type	3D, technical, graphic, corporate identity
Product type	Description, e.g. solid, liquid, powder, granules
Status	New product, new brand, brand leader, own label
Design requirements	Revolutionary, evolutionary, range extension, 'me too'
Sales	Countries of sale, markets, points of sale
Environmental concerns	Critical, noncritical; recycling, reuse, disposal

All changes to packaging, whether technical or graphic, have cost and practicality implications. It is of no value to develop packaging which is going to be too expensive to produce, or the production of which is going to be too difficult to control. In addition to design skills, there has to be a strong practical core to any good design studio. The following design-consultancy selection checklist identifies the areas that must be considered.

- *Creativity:* This is demonstrated by the agency's previous studies.
- *Technical ability:* The client must check on the agency's knowledge of materials and processes, its knowledge of

suppliers, its selection of equipment, its knowledge of distribution, warehousing and transport, and its awareness of the cost implications of design.

- *Artwork and print expertise:* The client must check on the agency's ability to produce artwork and use the correct methods, its knowledge of print processes and origination, and its knowledge of overprinting, bar codes and security coding.
- *Legal expertise:* The client must check on the agency's knowledge of national, international, EC and hazardous-goods law.
- *Environmental ability:* The client must check on the agency's track record, and its knowledge of regulations, codes of practice, and symbols.
- *Planning ability:* The client must check on the agency's scheduling abilities and its knowledge of planning methods.
- *Administration:* The client must know the agency's costing structure, fees, expenses and billing procedure.
- *Staffing:* The client must be acquainted with the agency's staffing structure, and the numbers and quality of the designers, sales staff, administration staff, seniors, juniors, account handlers, and contacts.
- *Facilities:* The client must check on the agency's equipment, communications, workshop facilities and computers.
- *Other areas:* The client must check on the agency's previous and current projects, any clash of interests, the agency's record keeping, whether any work will be subcontracted, the agency's film-making, photography, illustration, and model-making skills (as relevant), the agency's presentation skills, any alliances with other groups, trade bodies, organizations and universities, and overseas work done, and languages spoken.

A client will primarily want to employ a design consultancy for its creativity. Normally, companies turn to outside agencies for the vital spark of creative flair which is far more likely to come from a fresh and open approach than from within the company's own organization. Design is also concerned with solving problems. Again, specialist skills may be required that are not available inside the client company.

Judging creativity is a personal matter, and there are no guidelines. People either like and admire what they see, or they do not. Creative solutions to design problems do not need to be bizarre to be creative. Of more importance is the rationale behind the solution, and the thought which has been given to a design that meets the brief. Only on very few occasions will a design be so outstanding that it takes people's breath away. Unlike art, design is a compromise, and very few compromises are truly outstanding.

If creativity is a matter of personal taste, then there are other criteria which are much easier to quantify. There has to be a level of technical expertise within any design studio, even in one that is only involved with graphics. When selecting a design agency, one should find out what level of technical skills are available, and, wherever possible, meet the individuals who will do the work. The depth of the knowledge required will depend on the nature of the project, but, at a minimum, the team must have a knowledge of suppliers, machinery, and packaging materials and processes. Without this, and a knowledge of the cost implications of any recommendations made, the results may be useless. There is absolutely no point in spending considerable sums of money, and ending up with designs which look great but are not practical.

For any project involving graphic design, the design studio must be able to offer expertise on all aspects of print production. This should be thoroughly investigated, because many projects, while they have successful design elements, fail when translated into print. There must be familiarity with all the main print processes, so that what is possible and what is not can be established. This is a very specialist area, but it is vital if a studio is to see a job through from start to finish with a minimum of delays and problems. Most studios, but not all, produce their own artwork. The ability to produce artwork and total familiarity with electronic artwork generation is essential. This must include photography, illustrations, and the use of colour. Artwork inevitably becomes an issue at the end of a project, and so one must identify at what point the studio wishes to involve its print expert. It should be fairly early on in the study, and certainly before a first-stage presentation. There are many instances of a client becoming enthusiastic about a particular design route, and then being disappointed with the printed results. It is the studio's job to ensure that the final printed form of every design it recommends is as expected. However, this is no easy task. The client must be confident in the studio's assurances that the design will print as expected, because it is often too expensive or impractical to carry out full-scale proofing. Increasingly, when proofs are electronically generated, some aspects of the final design cannot be seen. Often, the colour may not be quite right or the halftones may not be accurately reproduced. This is a difficult area for the client, who may have to approve designs on the basis of a proof that does not represent the finished job. Hence it is necessary to establish in any dialogue with a potential design studio its level of competence and the entry point of the print department into any design study.

It is unrealistic to expect design studios to be experts in the legal aspects of packaging. They will assume that all the copy is legally correct as supplied by the client unless they have been specifically alerted to the need for legal checking. If the project is likely to have a significant legal aspect, perhaps because of national, international or specific regulations, such as those concerned with the transport of hazardous goods, then the competence of the studio in this area must be established right at the start.

In addition to practical design skills, a good studio must also have administrative skills. The client must know how much the project is going to cost, and when the results will be available. The client must beware of the consultancy that is unable to provide a firm quotation. Most studios cost out their work on an hourly billing rate. It is quite likely that it will not know exactly how long a project will take, but it will use its experience of similar projects as a basis. The client must make sure that it is understood that the agreed fee will be firm, even if the studio spends many more hours on it than it budgeted.

Finally, it is probable that a close working relationship will be established between the client and studio for the duration of the project. The client's representative must judge how this will work out on a person-to-person level. Will the studio, for example, be prepared to visit the relevant works and how they get on with the client's production manager? There are still a few consultancies that are not prepared to send their staff far, or that select them for such a mission on the basis of their ability to get on with people rather than their design skills. These consultancies do not last. It is worth establishing which individuals will work on the project, and who will have the overall responsibility for coordinating all the studio's activities, and keeping to budgets and meeting deadlines.

The structured approach of the checklist allows comparisons to be made between agencies. The studios' portfolios can then be examined in detail. Any conflict of interest should be declared and resolved if possible. The vital area which remains to be discussed is that of cost.

BUDGETING

One of the most difficult areas of packaging design is that of assigning and agreeing budgets. Budgets must be appropriate to the task in hand, and it has to be recognized from the outset that design is not cheap. The aim has to be to get the best value for money, and that process starts by identifying exactly what is required and when it will be needed. As with all purchasing operations, what is required is the right quality at the right price at the right time. If, as sometimes happens, a packaging study becomes a rushed exercise, it costs more, and it may also be unsatisfactory. It is better to plan in advance, which allows one to shop around to establish costs, timing and the optimum design source.

When an outside design consultancy is used, its attitude to costing out the project will depend on a number of factors. The consultants may feel, for example, that it is worth quoting a keen price just to add a prestigious name to their portfolio. It may also be the sort of work that they are keen to accept to broaden their design capabilities. Most consultants have an eye to further work, and are prepared to quote lower prices if they feel that there is a chance of developing new long-term core business. It would be wrong for a potential client not to be honest in making an approach along these lines. It is as well to be absolutely frank about the possibilities of further work, for example, rather than trying to drive prices down on the basis of false hopes. The design world is a small one, and word soon gets around. While studios have slightly differing methods of pricing jobs, most attempt to quantify the project in terms of hours, and then apply either a flat rate per hour or the billing rates of the particular designers allocated to the project. Only then do the other factors mentioned above come into play, when the studio management takes a commercial decision about the desirability of obtaining the work. It is notoriously difficult for the head of a creative team to know how long it will take to crack a particular design problem. On occasion, the first design route tried turns out to be the best solution. There then follows further work to prove that this is in fact the case. From the designer's viewpoint of believing strongly in the first piece of work, the subsequent investigations seem superfluous, as if work is being done for the sake of showing the client that all the routes have been fully covered. This problem can be avoided by assigning a team to the study, which allows a number of design directions to emerge early on. The most promising is then selected for development. Even so, the calculation of the time involved is a guess that is modified by previous experience. No more than a handful of top design companies specialize in packaging, and their fees are broadly similar. Clearly, however, commercial factors are more often influencing decisions, distorting the true value of the work in favour of lower prices quoted for strategic reasons. There can thus be wide price variations between agencies. It is not uncommon for a consultancy to agree a fee which it will honour, but it will carry out many more hours of work than are being paid for to achieve a result that is good enough to persuade the client to award it further stages of the project or further projects in the future. This is particularly true of the first stage of a study, where the major creative effort is concentrated. Studios may 'overspend' by 1000% or more, hoping to pull back some of the cost in future stages.

This all sounds like good news for the prospective client. Nobody wants to pay unnecessarily large sums of money to design agencies, or any other organization. There is a danger that budgeting will be focused on cost, to the detriment of quality and timing requirements. The lowest-cost option is not always the best one, but, provided that the selection process has been carefully carried out, low costs are welcome. It must be remembered, however, that studios will try to recoup any artificially low costs in other ways.

The above discussion relates to design. Consultancies tend to invoice for costs and expenses separately. When these cover travel, administration and materials only, they should be unlikely to exceed 10% of the fees, as a rough guide. However, particularly at later stages, where artwork, print trials and proofing are involved, costs and expenses may exceed the quoted fee. When studios are reluctant to be firm about the expected level of expenses, it is as well to agree an inclusive cost which cannot be exceeded without prior agreement. An arrangement should be made that triggers a warning as soon as expenses reach an agreed level.

If it is possible within the organizational structure of the client company, price reductions may be obtained for the prompt settlement of accounts. Many consultancies have to chase clients for money. This is an embarrassment for both parties. The client may feel that he/she has been overcharged and be slow to clear payment, and the studio may be reluctant to hound a client who may yet provide further projects. The method of payment should be agreed at the start of a project, in terms of the maximum total cost authorized, and how and when payment will be made. This seems straightforward, but there can be tremendous problems if it is not addressed early enough. Typically, problems arise in multinational companies, where the work may be authorized by one department, or one country, and payment is made by another division, perhaps in a different country. Currency can also be a problem. Payment in ECUS to one consultancy created rather a stir at their local high-street bank.

Using a strategic approach to packaging design, a consultancy should be able to indicate the number and nature of the stages that it envisages will be required for a study. As design is such a fluid business, it is difficult for a consultancy to be accurate about the price of each stage. The first stage, however, should be firmly quoted for, and this should be the largest portion of the total price in terms of fees, but not in terms of costs and expenses.

There is no reason why studios should not be able to estimate, at least, the prices for additional stages. They realize that client companies have budgets, and that some overall price will be required to obtain approval. Projects vary, but, as an example, a typical price breakdown for the graphic design of a biscuit carton is shown in Table 2.2.

Table 2.2 shows that 40% of the total fees are earned during the first stage, while 60% of the costs and expenses are not incurred until the fourth stage. The danger here is that, right at the end of the project, artwork costs can suddenly put the project overbudget. This possibility needs to be firmly controlled from the outset.

Some design companies may also charge an additional rate for the use of computer time. Originally, this was done for those clients who simply wished to use a designer working on a computer as directed by the client. This can often be a quick and effective method of investigating design concepts. To continue charging computer time on project work, however, does not seem to be justified. After all, computers have now become simply another design tool, in the same way as airbrushes and

Table 2.2 Price breakdown for graphic design of business carton

Stage	Activities	Fees, %	Costs and expenses, %
Stage 1	Store survey Purchase of samples Design display panel Experiment with type styles, backgrounds, colours, photography and illustrations Interim meeting Further development Presentation	40	20
Stage 2	Further development	20	10
Stage 3	Extension to other panels	20	10
Stage 4	Artwork	20	60

colour photocopiers are part of the normal equipment used by studios. Attempts by studios to recoup their capital expenditure on computer equipment in this way should be resisted.

PROPOSAL DOCUMENTS

When more than one design consultancy quotes for a job, how they respond will have a bearing on which one is selected. The consultancies will listen to or read the brief, discuss the matter internally, and formulate their response, which is usually formally set out in a form of proposal document. Most consultancies wish to present their proposals in person, either by visiting the potential client, or by inviting the client to the consultancy. In this way, they hope to talk through their concept of the project, how it will be conducted, how long it will take, and how much it will cost. They will also submit a document that summarizes this for subsequent perusal by the client.

Proposal documents vary from the simple to the elaborate, but they should all contain the following:

- An outline of the brief that is not just a playback of what the client has told the consultancy, but that demonstrates an underlying comprehension of the project in hand.
- Clearly stated packaging objectives.
- A description of how the consultancy is planning to approach the study. This should be detailed for the first stage, listing each scheduled activity. Subsequent stages can be less detailed, but there must be a logical structure for the complete project.
- A statement of how much it would cost the client. A firm fee should be quoted for the first stage. Costs and expenses should either be included in the fee, or estimated separately. The prices of subsequent stages should be estimated, and there should be an estimated price for the total project that includes estimates of all the costs and expenses.
- A statement of how long the project will take. A detailed plan should be presented for the first stage, together with estimated timeframes for subsequent work. A properly presented project schedule is a bonus that should count in the favour of any company that goes to this trouble and has this expertise.

Proposal documents can be elaborate affairs that are slickly prepared and incorporate glossy photographs of the consultancy's previous work. The way in which the document is prepared provides a clue to the nature of the consultancy. Creative companies adopt a more creative style, and the more pragmatic consultancies have a more technical approach to the document. Documents should not be too elaborate; if the client employs the consultancy, he or she will pay for the preparation of the proposal documents in some way.

It is, however, useful to have details of previous case histories, as this indicates the type of work undertaken. It will not be possible to judge the results accurately if not much of a brief accompanies the description, but the client can note elements of the style, and see what the ratios are of graphics and 3D work, new brands and established brands, specialist areas etc.

Some consultancies also list the principal contacts within the studio that the client is most likely to work with. This is useful, but the client may want to meet the contacts as well as read about them.

Packaging can be one of the most important areas of a company's expenditure, and it is important that the right people are chosen for the right task. They may not be the cheapest, but they should be the best. The client will employ them for their creative ability and their analysis of what is right for the product. It may be necessary to delve beneath the suits and expensive cars of the consultancy's marketing team to find out just how good the consultancy is. The proposal document is the consultancy's marketing tool, and, just as a car is not bought on the basis of the brochure and sales talk alone, the proposal should not be the deciding factor. The client should meet the team, question the work, and talk to other companies who have used them. It is essential to choose the right people.

FREELANCE DESIGNERS

Many high-calibre designers work on a freelance basis for one reason or another. They tend to specialize, and they cannot offer the breadth of services provided under the umbrella of a consultancy. Consultancies use freelance labour as a matter of course across the whole spectrum of design work. It is much rarer for a manufacturing company to employ freelance designers directly. This may be because there are no communications with freelance designers within the company organization. After all, where does one look when one requires a few days of design time for a particular project? There are agencies who maintain a register of freelance designers, and publications in which designers list their specializations.

Although tracking down the right designer may be difficult, there are benefits for the client in having a specialist dedicated to the particular task. Costs are directly related to time spent, and freelance costs are considerably lower than the prices of a consultancy with high overheads to recover.

UNIVERSITIES AND COLLEGES

The fact is often overlooked that universities and colleges are a design resource. Not only do they contain keen young designers in the making, but also experienced teaching staff. Many lecturers have moved from industry to education, and most still continue to work on a freelance basis outside the campus. This is important in that it not only augments their salaries, but also ensures that they do not become so engrossed in academic issues that they become out of touch with commercial design practice.

For small companies with limited resources or large companies who wish to establish links with universities, suitably directed projects can yield interesting results. They generate an entirely fresh approach to design that is largely unencumbered by preconception. Caution is required, however, as some concepts thus generated may be impractical, as the student designers are unlikely to have been exposed to industrial manufacturing processes.

The stage at which educational establishments in the UK charge for their work has not yet been reached, although this cannot be far off if current UK Government policy is pursued. Colleges seek to strengthen their links with industry, and, in particular, to feed their students with 'live' projects, and in return they generate design concepts. This process usually entails a manufacturing company briefing a group of students, reviewing progress at interim stages and assessing the results. For the benefit of the students, the project should be conducted in exactly the same way as it would be with a consultancy, including formal presentation of design work. A commitment of marketing time is required which can be considerable. It is not unreasonable, however, for those making a living from packaging to put a little back into the profession which has supported them over the years. The results may also have a spark of creativity which would have been difficult to find elsewhere.

PACKAGING SUPPLIERS

Some 25 years ago, before specialist packaging-design consultancies became established in the UK, it was normal for suppliers to offer assistance with design. This was part of the supplier/customer relationship, and, consequently, no direct charge was made for this service. Today, despite the growth in the numbers of design consultancies, many suppliers still offer a design service as part of their normal conduct of business. For valued customers, these services are 'free', although they obviously have to be paid for somewhere along the line. The level of design assistance has increased with the advent of electronic design systems that are aligned with production methods. Both glass-container and plastics-container manufacturers have been quick to grasp the opportunities afforded by new technology, which allows reduced leadtimes from design to container production, and provides the benefit of being able to show the customer 3D visuals of the finished pack on screen. Similarly, carton manufacturers and label printers have embraced computer equipment and software that has both production and design capabilities. This certainly makes suppliers an attractive source of design, and, in addition, whatever they design should not suffer from manufacturing problems, as might be the case with designs generated by a consultancy.

Not unnaturally, suppliers are biased towards their own products, and one cannot expect otherwise. There is little point in providing a supplier with the same type of brief that a design consultancy requires. A supplier's brief must be focused on the client's requirements. Suppliers do not have facilities for researching markets or analysing problems. Their strengths lie in knowing what can be produced and the cost implications of design changes. Generally, suppliers opt for designs that do not raise many problems in production. While this is a benefit, it is also a drawback, as the design work then tends to be conservative rather than to push back the frontiers of technology. An increasing number of suppliers, however, employ staff with a design rather than an engineering background, and there is no doubt that the role of consultancies will be challenged by enlightened suppliers. Suppliers have the ability to invest large sums in capital equipment, and some exceptionally competent design work is now being done by suppliers.

If a supplier is to be considered a design source, the client's marketing team need to be very specific about their needs. Alternatively, if a consultancy is involved, there is a definite advantage in having the supplier on the design team. Consultancies eventually have to discuss their proposals with suppliers in any case, to establish practicalities and costs. In some projects in which the design outcome in terms of container type is already known, it makes sense to bring in the supplier at an early stage.

3
Preparing the brief

One of the keys to packaging-design success is the preparation of an accurate brief. Many projects falter or fail because the brief has not been properly thought through, or because it changes throughout the project lifecycle.

The purposes of the brief are as follows:

- *Crystallization of thinking:* The process of writing the brief provides an opportunity to thoroughly think through the potential of new or revised packaging, and this can be an illuminating exercise. The relationship between the product and the pack and the point-of-sale requirements move sharply into focus.
- *Identification of objectives and constraints:* The discipline of writing a brief ensures that the objectives (i.e. what exactly needs to be achieved) are given proper consideration. There are also constraints, perhaps because of the nature of the product, or its production, cost or display requirements. Writing these down clarifies the purpose and limitations at the outset.
- *Communication:* The client's thinking has to be clearly communicated to others within the client organization and/or outside agencies. A brief is a communications document which forms a datum point from which to work for all concerned with the project. It is referred to as the project progresses to check whether designs are on course, and to measure success. Requirements and constraints specified in the brief dictate the direction of the project. It is impor-

tant that care is taken to ensure that the brief is accurate. The brief must be written with its target audience in mind. Terminology and in-house jargon must be explained where confusion could arise. Communication is the understanding of ideas expressed by the originator by the recipient.

- *Yardstick for cost comparison:* A brief provides a consistent standard which can be used to assess competitive quotes for the work. This ensures that a cost-effective approach is taken to tender assessment.

A brief also has to be realistic in its demands, and not be so all-encompassing as to be unworkable. The client may wish to reach for the stars, but they must be within grasp. Faced with the task of choosing a new car, most people are prearmed with a list of needs and requirements. They may seek performance, styling, imagery, engineering excellence, after-sales service, and an appropriate size and price. There will be a combination of requirements, depending on the person's circumstances. The Porsche may be what the consumer would actually like, but he or she may settle for an estate car which accommodates the family, the dog and the budget. A compromise is accepted because of practical considerations. How much more difficult it would be if a consumer had to design his or her own car. Starting from a blank sheet of paper, it would be difficult to end up with just one vehicle which satisfied all his or her needs. Some priorities would have to be established, and some desirables either incorporated or omitted, depending on circumstances.

Many packaging projects entail the development of a new pack form for an existing product, or a totally new product. The brief must be clear about the approach required, which will either be evolutionary, based on existing pack or brand strengths, or revolutionary, breaking the mould. The design approach will be quite different for each of these categories, and the costs involved will be dependent on the amount of work required.

The use of a checklist helps in the creation of an accurate brief. The checklist below (which is based on the checklist for a packaging brief given in Reference 1) is comprehensive, and it may not all be applicable to every study. Nevertheless, all the points should be considered before they are accepted as relevant or dismissed as inappropriate. One of the key functions of the brief is to crystallize thinking, and so do not dismiss checklist items before giving them proper consideration. It is tempting to skim across points that are difficult to resolve, but this should be firmly resisted. Even when a particular point of difficulty is peripheral to the main thrust of the study, it is often worthwhile to resolve the questions that arise, if only because the process allows the development of a deeper personal insight into the product and its packaging. The brief exists to communicate with others who may be less familiar with the product type and market than the client. What is second nature and very obvious to the client may not be picked up by an outside agency unless it is incorporated in the brief.

- *Market:*
  - size, volume and value now, future trends,
  - brands and brand share,
  - seasonality,
  - regional, national and international factors,
  - distribution: wholesale and retail,
  - purchaser, consumer and/or end-user profiles, and purchasing decision makers:
    - age,
    - sex,
    - socioeconomic groups,
    - special characteristics.
- *Product:*
  - nature of product,
  - history,

- brand share,
- competing products,
- brand loyalty,
- purchase motivation,
- point of sale and positioning on shelf,
- copy strategy,
- planned advertising and promotions,
- sacred cows.

- *Design objectives:*

  - new product,
  - redesign or update,
  - evolution or revolution,
  - range integration or extension.

- *Consumer requirements:*

  - standout,
  - quantity, size, volume and/or weight,
  - pack style appropriate to product,
  - size and value-for-money impressions,
  - compliance with existing market segment,
  - associated shapes, colours and pack styles,
  - inspection and handling prior to sale,
  - copy and instructions for use,
  - benefits and added-value features,
  - portability and ease of carrying,
  - storage and in-use life,
  - opening, closing and resealing,
  - dispensing and/or pouring,
  - safety: protection against product hazards,
  - after use,
  - tamper evidence and child resistance,
  - environmental assessment.

- *Product requirements:*

  - compatibility of materials and product,
  - product life and deterioration characteristics,
  - shelf life,
  - product sizes, weights and volumes,
  - protection against

    - liquids and moisture,
    - gases and odours,
    - microbiological contamination,
    - temperature extremes,
    - infestation,
    - ultraviolet light and artificial light,
    - mechanical damage through handling, transit and/or storage.

- *Production requirements:*

  - need to use existing plant,
  - machinery and lines available,
  - opportunity to introduce new processes and machinery,
  - in-house facilities: materials, components and/or printing,
  - cost breakdown:

    - prefilling,
    - filling,
    - closing or sealing,
    - weighing,
    - labelling and printing,
    - inspection,
    - collation and overpacking.

- *Purchasing requirements:*

  - tied or favoured suppliers,

- pack quantity and cost data,
- contract packing.

- *Warehousing requirements:*
  - storage methods,
  - pallet type and size,
  - stack heights and weights,
  - storage times,
  - opportunities for standardization of shipping cases,
  - optimum pallet-loading patterns,
  - pallet and outer identification system.

- *Transport and distribution requirements:*
  - distribution methods:
    - mail order,
    - wholesaler,
    - retailer,
    - direct,
  - storage times, at each stage,
  - need for divisibility, at each stage,
  - vehicle-loading methods:
    - mixed goods,
    - cage pallets,
    - side loading or end loading,
  - pack/pallet identification,
  - security and antipilferage methods.

- *Merchandising requirements:*
  - size compatibility with fixtures,
  - use of planograms,

  - back-card sizes and Euroslot configurations,
  - need for tactile product examination on display,
  - need for price-marking area,
  - position of display:
    - shelf,
    - gondola end,
    - counter,
    - dispenser,
    - checkout,
    - window,
  - position relative to eye level (above or below),
  - display by product type and brand,
  - panel orientation: front, back, sides, top, bottom,
  - lighting levels at point of sale.

- *Print requirements:*
  - number of colours and varnish,
  - available print area,
  - print method,
  - materials,
  - quantity and size of print run,
  - additional future copy and promotional flashes,
  - multilingual copy,
  - bar codes, security codes and batch codes,
  - area for overprint data,
  - special requirements for top, bottom, sides, ends,
  - tie in with advertising,
  - costs.

- *Legal requirements:*
  - national, EC and international compliance,
  - special attention to

    - weights and measures,
    - ingredients: copy, size and order,
    - manufacturer's name and address,
    - misleading illustrations,
    - product claims,
    - hazardous-goods notification requirements,
    - symbols,
    - materials and inks restrictions.

- *Environmental requirements:*

  - minimum packaging and no overpacking,
  - minimum-energy utilization,
  - recycling and reuse considerations,
  - disposal,
  - correct use of symbols,
  - avoidance of use of mixed materials,
  - no claims of environmental superiority.

- *Administration:*

  - timing requirements,
  - cost budgets,
  - research requirements, existing and planned,
  - implications for advertising,
  - associated studies, for example shipping cases and point-of-sale material.

If all the points on the above checklist are considered, and detailed descriptions of each committed to paper, the brief will not only be comprehensive, but likely to run to many pages. This is fine for a brand manager who is familiar with every nuance of the market, and the brief will probably be an excellent reference document. For designers, however, who tend to think in pictures, it may be too detailed. Once the facts have been listed, it is often worth distilling the contents into a more concise, focused format for presentation to design agencies.

Alternatively, the design agency may summarize the brief itself, although there is then the danger that it may emphasize some less-significant issues. Ideally, the detailed brief should be a reference document that is available to designers, and that helps in the clarification of queries. The short version is then the working brief or summary, and it lists the key points that are vital to the study. Whenever possible, a verbal briefing should be given to the design team, rather than a written document being relied upon. Product samples, competing products, and pack forms that are admired and detested should be shown to illustrate points during the briefing session. Most designers respond far more to objects or photographs than to text. A table covered in packs and products not only illustrates points in the brief, but also starts designers thinking immediately about potential solutions. A question-and-answer session at the briefing is a valuable start to a project, and can prevent design work taking off in the wrong direction. Some design houses like to produce their own version of the brief in the form of a worksheet, or a summary included in the proposal document. There is some validity to this approach, as it should demonstrate the design company's ability to understand the brief and interpret the salient points accurately. Both worksheets and proposals should break down the project into stages, and provide costs and timing estimates.

In writing a brief, the first element should be background information about the market, product and current pack.

MARKET

The broad picture is considered below of the market within which the product will be sold, its volume and value, how it is fragmented among brands, and what the brand shares are. It is also necessary to predict how the market will change in the future, and what trends are likely to lead to change.

Some markets are seasonal, for example that for chocolate Easter eggs. However, we may wish to question whether the seasonal presumption is really valid. Cadbury did this, and it now successfully sells chocolate cream eggs all the year round. Markets may also be regional, with products appealing only to particular parts of a country. This may be the current position, but does it really have to be the future one? The mobility of labour, recreational travel, and advertising through national media (particularly television) mean that there is increasingly less emphasis on regional boundaries. The current position and future trends must be considered. Where markets span national frontiers, the requirements of individual nations are similarly becoming blurred. Within continental Europe, where television advertising crosses boundaries and language barriers, there may be significant benefits in producing one standard of packaging that is suitable for all. When a brief is prepared, the opportunity arises to question the current market position in addition to simply recording it. When new or revised packaging is being considered, there may be opportunities to expand the brand into new markets.

A brief must indicate the importance of wholesale distribution, if this is a significant factor in sales. With some products, the ability to sell wholesale can be as important as retail sales, with implications for the intermediate packaging. A breakdown of both wholesale and retail distribution may be required.

Another key issue in packaging design is that of the consumer. One needs to be clear about who is making the purchasing decision, and who is influencing the purchase. While purchasers of breakfast cereals are likely to be females aged between 20 and 45, children may be the influencers. The pack may have to reflect the child's values rather than the mother's, although it may be necessary to take into account the mother's concern to purchase products which are nutritious for and appetizing to children, and offer good value for money. The purchaser, consumer and/or end-user profiles need particularly close attention. Designers focus closely on packaging designs which appeal to the target audience. If this element is wrong, the study will not provide the right answers.

PRODUCT

No packaging project can be successful unless the product is understood. Later, technical details about the product will be needed, but the background section of the brief requires an outline of the nature of the product, how it is used, and what competition it will face.

A brief history of the product is useful to the designer as a backdrop onto which new ideas can be projected. The brand share of the product (both the current share, if the product already exists, and the forecast share) must be documented.

Brand loyalty, or the major motivation for the brand's purchase, requires analysis. Is it thought that the brand is purchased because it is trusted as an 'old friend', or because it offers something that competitors do not, or is simply good value for money? Designers want to know what it is about the product that can make it stand out from the rest. If competing products exist, the design team will need to see them. The brief should outline the advantages and disadvantages of the client's brand in comparison with those of the competition. This should be in terms of both product values and packaging styles.

The point-of-sale positioning is critical in packaging design. It is here that the designer will fight hardest to achieve the best standout against the competition. The marketing arm of the team has a responsibility to identify where the product is likely to be placed in each type of sales outlet that the product is going to be placed in. Later, the design team will visit outlets to see for themselves whether the client's estimate is correct, but a description of outlet types and point-of-sale positionings will suffice in this part of the brief.

If any copy strategy has been preplanned, this should be documented. Advertising and future promotional plans must also be considered. Many packs have to allow space for later promotional use, and it is as well to make sure at the start of a project that the design team understand the need to reserve areas of the pack for this purpose. With many types of product, the pack forms an important part of the advertising strategy. Often, the pack itself features strongly in an advertising campaign. This need not necessarily be a constraint to pack design, but it must be borne in mind.

There can be 'sacred cows' in packaging design, which are areas of which the designer should beware. If there are devices, graphic or otherwise, which are so sacrosanct as to be mandatory, the design team must be forewarned. Often, such a device is a company logo, which, however unpleasing, was originated by the chairman's wife for example, or is part of the company's illustrious past. The designer needs to be forewarned of circumstances like these so that he or she can avoid committing commercial suicide. For example, the silence in a room full of senior executives as they were told that their 100-year old logo was out of date was deafening.

When the background information has been considered and documented, the design objectives must be set. The approach differs according to the nature of the project, and to whether, for example, structural 3D design and/or graphic design is required. The starting point is the same, however: clearly stated packaging-design objectives.

PACKAGING-DESIGN OBJECTIVES

Convention dictates that all objectives should be prefaced with the word 'to'. While many conventions are simply standardized forms of communication, this one does have the additional merit of focusing attention on the key requirements.

Each project will have its own specific objectives, but in each it is necessary to consider whether the design is to be new and innovative, or an update of an existing design. Usually, updates are graphic projects only, perhaps being carried out when the pack has started to look dated, or it has been eclipsed by competing products with greater standout. Both graphic and 3D design can be used to create such evolutionary design work, although the freedoms afforded by 3D design lend themselves to revolutionary design work. The brief should allow for planned range extensions in the future, or be specific about the need to integrate the new design with current designs.

Whatever the nature of the brief, setting objectives is vital to the conduct of the study. It is a mistake to list too many objectives: the list needs to be brief and to the point. The objectives should not be so vague as to cloud the issues, because, if this happens, not only will the design work be too open-ended, but it will also be difficult to judge the success or otherwise of the results. It is no good defining an objective as, for example, 'to sell more products'. The objective has to be specific, and the target area for action has to be clearly indicated. 'To improve the standout against those of (named) competitors', for example, provides a goal to aim for and a tangible, even if arguable, measurement of success.

CONSUMER REQUIREMENTS

The inclusion of topics in the brief which are relevant to the consumer must be considered. Standout is one of these for every pack form targeted at the retail market. It is worth putting ourselves in the position of the consumer to determine and analyse what criteria are important. The checklist identifies several areas for consideration. Not all of these will be applicable to every study, but they should all be considered, and either rejected or identified as being of significance. Some of the less obvious items in the checklist should be explained. The need to handle goods prior to sale applies to that category of goods that consumers need to reassure themselves is right for the job in hand before purchase. Abrasive paper and abrasive discs are an example. Despite indications of the grit size on the pack, most DIY enthusiasts still want to feel the roughness. The product description of, say, '150 grade' is not a meaningful description for most consumers. It is not quite meaningless, but not much help either. The marketing executive has to recognize this, and brief the design team accordingly. It is then a design responsibility to produce a pack design which allows presale examination while still maintaining all the other functions required by the packaging.

Even in areas in which there is no identified need for a particular pack feature, unlike in the abrasive example, it is worthwhile trying to determine whether benefits could be achieved by incorporating something useful. Anything which makes the product easier to use, or easier to store or carry, provides a greater incentive to purchase. If the feature is obvious prior to purchase, so much the better. Subtle features that are discovered after purchase are still a positive benefit in terms of consumer satisfaction with the product/pack, and are likely to lead to repeat purchases.

One should also consider how the pack will be perceived by the consumer from an environmental viewpoint. There must be no attempt to fool consumers into thinking that they are buying environmental superiority. If, however, a real consideration of environmental issues can be demonstrated by the manufacturer, consumers are likely to respond positively.

Manufacturers put themselves in the position of being a consumer all too rarely. For years, consumers have struggled to open packs while manufacturers either elect not to listen to their complaints, or deny the problem exists. This attitude has little place in a competitive environment in which the consumer can vote for the pack and product which meets his or her needs. This must be the client's pack!

TECHNICAL REQUIREMENTS

Sadly, in some organizations, marketing staff move between brands with such rapidity that they never have time to become familiar with the products' more technical aspects. It is both fascinating and illuminating to spend time finding out how a product is made and packed. This is necessary for this part of the brief, as much of the information is essential to the design team.

PRODUCT REQUIREMENTS

When a new pack form is being considered that perhaps requires a radical change in materials, it is essential to know what conditions have to be met to ensure that the pack will perform efficiently throughout the product lifecycle from production to end use. The product-formulation chemist or the department responsible for product development is the source of this information. When new markets are sought, for example in hotter, more humid, colder or more distant countries, the packaging-materials selection must take climatic conditions and additional handling into account. For example, a flexible polyethylene spout on a new design of oil bottle worked very well in the UK, but snapped off in use in the much colder conditions of Scandinavian winters.

The need for compatibility between pack and product may seem to be an obvious requirement, and yet it is vital to ensure it. Suppliers can provide data about compatibilities with specific chemicals for most materials, but possibly not for complex products. Some testing may be required if there are any doubts, and a time adjustment must be made to the product launch plan to accommodate this phase. There must also be a budgetary adjustment to cover testing costs. A typical requirement is long-term product stability in the conditions expected in the country of sale. Specialist testing facilities may be needed to ensure that the pack will perform satisfactorily over its lifespan in terms of moisture-vapour permeability through the pack and taints from materials. Food products containing fats are particularly prone to picking up odours or flavours from packaging, chocolate being a prime example. Some products may also adversely affect pack strength. Polyethylene bottles containing oils or solvent-based products often exhibit 'panelling', where the container gradually deforms over a period of time owing to the product migrating through the container walls.

Every pack will be subjected to the rigours of handling, from the moment of filling to final disposal. Experience with current packs, if there are current packs, will provide a background to the level of mechanical hazards with which the pack will have to cope. Any new or modified pack form needs to be assessed for compatibility with the distribution system envisaged. This assessment can take the form of transit testing using preproduction pack samples fed through the normal distribution channels, or simulated transit tests can be organized using a specialist testing house. In either case, a problem may be that of being able to produce packs at an early enough stage that give results that will be in line with the packs produced by the final production line. Often, preproduction samples are not representative of the final quality, and, if one waits until production packs are available, it can be too late for adjustments to be made. The designer has to use his or her judgment to establish a suitable pack specification that can later be confirmed by testing. With plastics containers, for example, the tooling will be produced before tests can be conducted. Although it may be possible for the pack weight to be adjusted, there is then little further scope for modification. Increasing the pack weight also increases the pack cost, with a subsequent detrimental effect on the overall product costings. Usually, designers faced with this situation use a slightly higher specification than is required to allow a safety margin, from which they can decrease the specification in the light of experience, and hence reduce costs.

PRODUCTION REQUIREMENTS

The brief needs to take into account requirements to use existing facilities or to consider new machinery and plant. If capital expenditure is out of the question, any new pack design will have to work within the constraints of the current setup or with the minimum of modification. The brief must state clearly the scope that is available for the introduction of new plant, because otherwise designers may waste time and the client's budget looking at pack forms which are nonstarters. Conversely, if new thinking is to be part of the brief, and there are no constraints on plant, then the brief must say this, and the designers must include estimates for new plant to fill/pack any new designs.

PURCHASING REQUIREMENTS

Details of quantities are important in the investigation of new or revised pack forms. Cost data for current packaging must be included in the brief to allow cost comparisons to be made between existing and new packs. This should include call-off quantities as well as annual quantities. If contract packers are currently used, or may be considered for new pack forms, the brief should include the details. Some companies may use tied or favoured suppliers, and, of even greater significance, there may be suppliers who must definitely not be approached. This may be because of past bad experiences, or because of their known ties to competing product/brand manufacturers. Much of the detail is probably best left to designer–purchasing-department discussion, but it as well to alert the design team to the situation in the brief.

WAREHOUSING REQUIREMENTS

It can be of great importance that the design team knows what sort of warehousing facilities are to be used. For example, some companies operate a high-bay automated system, in which packs are automatically collated and stacked. The failure of a pack in this type of system can lead to a dangerous situation and/or the contamination of other packs, with expensive results. If palletization is a key issue, there is little point in spending time designing packs which do not offer improvements in palletizing efficiency. If there are opportunities for the standardization of shipping-case sizes, the designer should be alerted to the fact in the brief. This is also the case with existing in-house pallet-identification systems. These will probably have been developed and refined over years, and such a background should not be ignored by a competent design team. In some instances, order picking is highly automated, with bar-code variants being picked up by individual lift trucks. If the brief does not mention the detail, it should at least stress the need for the design team to speak to warehousing staff.

TRANSPORT AND DISTRIBUTION REQUIREMENTS

The brief must include a description of the distribution system that will be used. It does need to go into detail, and it should give the typical storage times at each stage, and indicate the need to subdivide collations of the product where appropriate. Some companies start with vehicle sizes, and work back to pack sizes to achieve maximum vehicle utilization. If this is an important issue, this should be stated in the brief. Even if it is not current practice, it should be considered, as cost savings could result. When a designer works on a unit pack, it is fairly easy for him or her to consider the unit in isolation, and not make connections with the subsequent operations. While the unit pack that is going to be sold may indeed represent the focus of any study, it is nevertheless a mistake to consider it in isolation. The packaging is part of a total system operating within a company, and each stage is important, offering its own economic opportunities. A brief which does not include transport and distribution is possibly missing a trick. It is an area where cost savings can be made and efficiencies improved, and the client must take time to consider it.

MERCHANDISING REQUIREMENTS

Merchandising is a key area in maximizing sales. While it may not be possible to persuade retailers to position the pack particularly advantageously on the shelf, positioning must be taken into account. Unless the pack complies with the demands of the retailers, and particularly the major multiples, the product may not be stocked, or the number of facings may be reduced. While it is possible for the designers to obtain this information, it is probably better if the information is provided by the client's marketing team, because they are familiar with the product category and the sales outlets. Later in the study, a retail-outlet survey will be recommended, so that the design team can see for themselves exactly what happens. Nevertheless, it is the marketing executive who is most likely to appreciate the finer nuances of product display, and pinpoint any opportunities or constraints. There have been many instances when this was not done, with resulting designer confusion. In one example, the designers thought that the pack would be displayed in a hanging format, but it was displayed on a gondola-end shelving unit. The lack of communication created wasted effort and unnecessary costs. Point-of-sale positioning will be checked out, but it should also be known and understood by the marketing staff, and the details made clear in the brief. Of course, what should take place often does not. It is not always the main panel of a pack which is displayed. However confident the client may be, a retail-outlet survey should be insisted on. It is not an expensive exercise, and it pays great dividends by revealing the real point-of-sale conditions (and this is often a revelation to marketing staff). For example, dog food is often displayed on a bottom shelf in supermarkets. At that level, the most obvious feature is the blank ends of the cans. In one example, valuable branding was achieved by printing the can ends. If the designers had concentrated on the main body of the can, the most important facing would have been overlooked. It was the store survey that revealed the actual location of the packs.

PRINT REQUIREMENTS

It is not always known before a study is carried out how many colours will be required to achieve the desired print results. When there are limitations because of, for example, print processes or cost constraints, these should be stated in the brief. Later, perhaps, the designers may challenge the brief, but if there are sound reasons for limiting the number of colours the brief should say so. The designers should be alerted to any future needs for promotional flashes or special-offer details on the pack. There are two schools of thought on this. Either the promotion should disrupt the main graphics, thereby achieving maximum impact, or the promotional flash should be incorporated seamlessly into the overall design so as not to ruin it. Thought needs to be given as to which approach is preferred. Security or bar codes must have areas of print reserved for them. Spaces must be allocated before the design work commences. Otherwise, a design which is admired may suffer because it does not work within a differently shaped panel that makes an allowance for promotional flashes. There are always cost implications in printing. If there are cost restraints, the brief must state them.

LEGAL REQUIREMENTS

Most design houses expect that all the copy supplied to them will have been legally cleared. Experience shows, however, that mistakes can occur, for example when the wrong size of type and/or graphics is used. Often, problems occur when the design studio is instructed to use illustrations or photographs that depict the product. It is essential that these should not be misleading. If whole fruits, for example, are shown on the pack, they must have absolute reference to the actual contents of the pack, and not be in any way misleading to the consumer.

It is unreasonable to expect design companies to have legal expertise, particularly in the form of a knowledge of legislation in other countries. The brief should either indicate a need for legal checking, and make an allowance for the employment of specialists, or the client company must undertake to carry out all legal clearances. It used to be standard practice for all artwork to be signed by a legal representative of the client company. With the advent of computerized artwork, this practice seems to have fallen into decline. It is important that some form of clearance is obtained by the consultancy, and that one individual is responsible for signing off artwork. Hazardous products need specialist attention, which is not always available within the design company or the client company. There are specialists who can advise on these at a cost. If the brief is for a hazardous product, the need for compliance with hazardous-product regulations should be highlighted.

Some countries also have specific regulations relating to the use of toxic substances in packaging, commonly specifying the avoidance of the use of heavy metals in inks and pigments. Designers and suppliers often have knowledge of such restrictions, but the brief should be clear about whose responsibility it is to check them out.

ENVIRONMENTAL REQUIREMENTS

Environmental issues are a very important area. Taking them into account should be an automatic design requirement, and every designer should try to uphold environmental standards. This should not be taken for granted, however, when the brief is prepared. The environmental performance of the product, its manufacture, and the pack must be considered. Undue emphasis should not be placed on the marketing opportunities afforded by compliance with environmental principles. The impact of the packaging on the environment should simply be reduced by making proper provision at the outset. Packaging should be simple and not overelaborate; its production should use the smallest possible amounts of energy and materials. The brief should contain a reference to environmental performance being one of the key criteria to be used when a pack is judged.

ADMINISTRATION

Administration is an area that is often overlooked in the excitement of drawing up a brief. It is also one of the areas that causes problems. Many projects falter when costs are exceeded or timing allocations are overrun. It is far better to establish the ground rules at the start, and avoid problems later.

The brief should be realistic about timings and costs. In writing the brief, some idea of the amount of work required and the timings involved will be gained. The need for research must be taken into account. This is most likely to be required after the end of the second stage. If research is required, time must be set aside and budgets allocated for the production of research material as well as the research itself. Often, the packaging is featured in advertising campaigns for the product, and this may impose deadlines on the packaging-development project. Allowances must be made for this, or the project may need to be telescoped halfway through. In fact, this often happens as the deadlines for advertising approach. Any associated studies, such as shipping-case development, must be included in the brief, or be the subject of a separate brief produced in tandem with the main brief.

The preparation of a comprehensive brief may initially be a daunting prospect for its writer. The brief will, however, repay the work invested in it by concentrating the efforts of those involved on the key areas, avoiding unnecessary time being spent on less-important aspects of the project. Preparing a brief is also one of the best ways of thoroughly analysing the product and the market, now and in the future. The client company should write a brief, and keep it accurate, to the point, and concise. It will then have a document that is of great value to its marketing staff and to outside agencies.

REFERENCE

1 *Checklist for a Package Design Brief* Siebert/Head, UK (Siebert/Head Ltd, 38 Hans Crescent, Knightsbridge, London, UK (Tel: 071 581 3839)) (the author of this book acknowledges permission to reproduce sections of this checklist)

4
Project planning

Major packaging projects can run for many months, especially if new tooling or extensive production trials are required. Even simpler projects are likely to need careful planning to meet deadlines imposed by product-launch dates, advertising campaigns, print trials and other activities. It is often at the final and critical stages of the project that panics, oversights and unbudgeted costs arise.

It pays dividends to plan the project as thoroughly as possible and agree the schedule with all involved at an early stage. Problems still arise, but a schedule can accommodate contingencies, or at least give an early warning of the need for action.

No two packaging projects are the same. There are parallels and similarities between projects, however, and this chapter looks at a simple study and more complex type of packaging study to illustrate features that are common to most design studies.

PROJECT OUTLINE

The starting point for all packaging projects is the brief. This provides the first indication of what is going to be involved, what the likely timescale will be, and how constraints may affect the size and timing of each stage. In writing the brief, some of these factors will be determined, but no detail will yet be available, as it is not yet known what direction the study will take, and what its implications will be.

One can, however, start to put together a project outline by considering the following factors:

- *imposed schedules:* seasonal requirements, competitor activity, new product launch,
- *design work:* timescale for depth of work required,
- *possible design outcomes:* off-the-shelf packaging, custom-made packaging, modified existing packaging,
- *market research:* if required,
- *production time:* existing plant, new plant, trials, stock build up.

Any imposed schedules will be known by the writer of the brief, and they may be such that the nature of the study is governed largely by one of the above factors. If the product were seasonal, for example, there might be a critical date by which the product had to be sold to the trade. This date might be only three months away. This deadline could obviously not be met if a radically new design were developed that involved either new tooling or new plant. In this example, the design study would have to eliminate any such contenders, and concentrate on less radical pack forms.

The design time required can be determined through discussion with the design source, or a fairly accurate estimate can at least be calculated.

The outcome of the design study will be unknown but, with a knowledge of the brief and preferably after discussions with the design source, the timescale can be estimated. For example, if any injection-moulded components are likely to be incorporated into the pack design, or the pack will be subject to any process which requires the initiation of tooling, the schedule must allow at least ten weeks for tool manufacture. On the other hand, if the designers are considering cartons (even a new style of carton), leadtimes can be considerably reduced. It is necessary to predict the longest and shortest expected project durations, taking into consideration the nature of the project, and its complexity. The designers, who may at this stage simply be quoting against the brief, should be able to provide guidance from past experience. As a rule of thumb, any project, such as the development of a new (blow-moulded) plastics container, involving tooling is unlikely to be completed in less than 12 months. Most projects do not have open-ended timescales, and so longer-term design options may not be possible. The designers must be alerted to this at the outset.

Should any market research be required, a considerable amount of time may need to be added to the outline plan. While it is usual for the actual research to take about one week, and the analysis of the results a further week, time must be allowed for the production of research material. This often involves model making or photography, and it can extend the timescale by another 2–3 weeks, even for a simple research programme. The logistics of delivering the research material to the test sites, particularly if these are abroad, should not be ignored.

An agreed design is a starting point for pack production. It is also the starting point for gearing up the filling and packing lines to accommodate the new pack style. If this is likely to involve capital expenditure on new plant, administration time may be required for approval by top management to be granted. New plant, depending on what is needed, may require a long leadtime, possibly of months. Pack styles which require such plant could be ruled out on the grounds of time requirements alone. There will also be a commissioning period during which any new equipment is set up and run. At this stage, pack samples must be available that represent the final pack specification so that the machinery and the packs can be checked for production acceptance. Finally, it will be necessary to build up stock and distribute it to sales outlets before launch.

With these and other considerations all being focused on a preagreed launch date, life can become very difficult. Project planning is essential for the control of each peripheral activity, and to ensure that, when possible, activities run in parallel to save time. The plan also needs to take staff holidays, public holidays and weekends into account. Design studios and marketing staff involved with the project may be quite willing to work around the clock. Most packaging suppliers are not, however, and it is quite common for a factory to shut down in the summer for two weeks. This could well be a critical period during the project. The plan must make allowances for production shutdowns. Executives go on holiday in the summer, and this is a problem if the executive who must sign off the artwork or approve a design modification is not in the office. Everyone involved in the packaging project, and particularly those key members who cannot delegate responsibility, should be on the circulation list for the project plan, and they must be alerted to the need for their presence at specific moments.

There must be some built-in flexibility to allow for contingencies and changes as the project progresses. The planning schedule cannot be unalterable. It will need to be continuously updated as more accurate data becomes available. There has to be a point at which commitments are made, particularly when an advertising campaign has been arranged for a specific launch date. Experience shows that most planned projects do meet their deadlines, although they often just scrape home with minimum stock on launch day because the schedule becomes increasingly punishing as the launch date approaches. It is preferable for one person, who can chase individuals or companies to maintain progress, to act as a project coordinator. This can be a thankless task, although it is made easier if it is ensured that everyone is aware of the power invested in the coordination role. If the coordinator is a member of the design consultancy, he or she can have difficulties in dealing with the client company and the suppliers. Instructions or requests for action from an outsider to these organizations may be met by a poor response. The authority of the coordinator must be recognized by all parties. Equally, it is desirable to select a coordinator who is tactful as well as insistant, so as not to alienate the parties involved. Every individual will have a part to play in the success of the project, and the coordinator must generate a team spirit in order to pursue the project goals.

The following is an example of how this approach works in practice. It is a simple study that is typical of the type handled by design consultancies on behalf of fast-moving consumer goods (FMCG) companies. It is followed by a more complex study in which an oil company redesigns its range of lubricants.

SIMPLE STUDY

In this study, the client company is a manufacturer of soups, and a brand leader within the low-fat, calorie-controlled soup market sector. It produces soups in cans, but sees the market being eroded by packet soups. A range of powdered low-calorie soups is being developed in response to this challenge, in a variety of flavours. The powdered-soup sector is further divided into markets for individual sachets and cartons containing multiple packets. It is this latter route that the company wishes to pursue. Manufacturing has been geared around the liquid product, but the company has a sister company which has facilities for packing powders on a form–fill–seal line, and which can also offer carton erection and filling. Because of pressures from competitors, and, more recently, own-label goods, the company wishes to make a swift entry into the market, using its brand name as a champion of accepted quality. It is accepted that contract packing will probably be required initially, but the company will later wish to consider an in-house packing operation. The pack formats used by competitors are already established; they all use cartoned sachets. The company is looking primarily for standout through graphics, but it is open to the suggestion of a 3D pack that might provide distinction on shelf. Whatever the pack form, it must carry graphics which not only highlight the brand name, but also indicate the calorie-controlled attribute; these must be adaptable across a range of flavours. Known for its quality liquid soups, the company believes that the brand name is strong enough both to sell in to the stores and to sell out to consumers. A television advertising campaign is planned to launch the new range of products. Five flavours are being considered for the launch, with further flavours and varieties to follow. The packaging objectives stated in the brief are as follows:

- to be distinctive on the shelf,
- to promote the brand,
- to indicate clearly the fact that the soup is 'calorie-controlled',
- to be adaptable across a range of product variants.

The timescale can be estimated by considering the project outline:

- *Imposed schedules:* There are none, other than the need to get into the market as swiftly as possible. There will be a need to link the packaging to an advertising campaign, as yet unscheduled.
- *Design work:* A design study is required to determine whether there are any pack forms which offer improved standout against competitors. This is expected to confirm the use of cartons, and the study is fairly low-key (timing estimate: one week). The graphic-design investigation is the main thrust of the project, and four weeks are allowed for this.
- *Possible outcome:* The outcome will probably be a sachet within a carton, with existing technology and contract packers being used.
- *Market research:* Research will have to be carried out to ensure that consumers will accept the new product range and link the established brand values of the existing canned products with a packet soup. Three weeks is estimated for this activity (UK only).
- *Production time:* It is expected that contract packers will be used who are already geared up to this pack format. The preparation of the artwork will take around four weeks, the pack production a further four weeks, the actual packing of the initial stock about three weeks, and distribution to selected outlets a further three weeks.

At this point, all that we have is a very rough estimate of timing, but already the scale is becoming evident. From the above, our project is unlikely to be completed in much under 22 weeks, or some six months after the design work is commissioned.

This is probably enough detail to enable decisions to be made at this stage about the viability of the project in terms of timing. The costs are not known, and nor are any design solutions, except for the assumption that the most likely, and probably the quickest, route into the market will be to produce sachets within a carton format. Approaches to design consultancies will probably be required for more detail to be added to the plan. Consultancies will be asked to provide a proposal document which outlines timing and costs that is based on the written brief and individual discussions. If the consultancy is familiar with strategic design practice, it will submit a proposal that indicates the activities that are likely to take place, the time that each will take, and the estimated associated costs. The client can then construct a more detailed timing plan in conjunction with the selected consultancy (see Figure 4.1).

For this study, the consultancy has elected to combine graphics and technical investigations into one plan, which shows a two-stage design investigation prior to research, with one final stage covering further development and production aspects. Secondary packaging is included in the plan as a separate but linked area.

The first stage includes a store survey, factory visit, and discussions with the advertising agency; these take place in parallel. The initial conceptual 3D and graphics work gets under way fairly soon afterwards. Although the marketing department is prepared to accept a pack similar to those of the competitors, they are also open to other formats, provided that they confer competitive edge in terms of standout.

The store survey reveals that all the competing packs use the same carton format and size, and are displayed by brand rather than variety. A display tray is used to collate the packs, with some brands having graphics on the lip of the tray. Signifi-

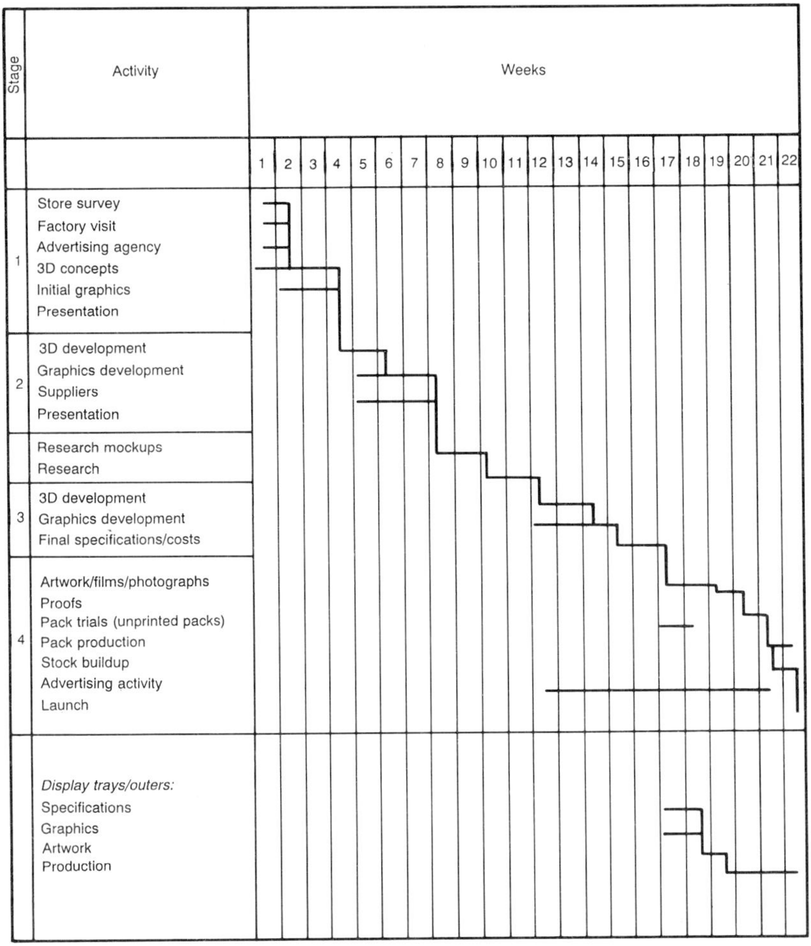

Figure 4.1 Project plan for simple case study

cantly for graphic design, this lip extends upwards by about 30 mm. Any text or graphic device printed at the bottom of the front face of the carton is hidden by the lip. Later, it will be important to ensure that this is taken into account, or that attention is focused on an alternative style of collation tray. One can be fairly accurate about the activities scheduled to take place during the first stage and the timings involved. The plan allows for a store survey in selected UK outlets, a factory visit, and discussions with the advertising agency; all these will take place in the first week of the project. Although the outcomes of these will not be known, the consultancy feels confident enough to also start the initial conceptual work on graphics and 3D design. This first stage is planned to take four weeks, with an interim review after two weeks, and a full-scale presentation at the end of the period. If the client agrees to this, the actual dates, times and locations must be set and met.

As the plan is projected further into the future, there is less certainty about timings. The outcome of the design work can only be predicted, and factors may emerge which change the pattern of events. The second stage entails a further week of 3D development, largely to take on board reaction to the recommendations resulting from the first-stage presentation, but also to prepare drawings and/or specifications and outline costs. Graphics development, which in the first stage is concentrated on one product variety with some exploratory work being done on range extensions, now starts to finalize the basic design, and extend it, in detail, to all five varieties. There remains the question of whether to use in-house packing or contract packers. Time is allowed for this to be resolved, and for suitable contract packers to be selected if necessary. Three weeks are allowed before the presentation that concludes the second stage.

It is usual to carry out any market research at this point. Although the nature of the necessary research is as yet unknown, two weeks have been set aside for the production of mockups of packs of the correct format that carry graphics for each product variant that show the main panel of the pack only. Two weeks are allowed for the research, with one week being allowed for research, and one week for the analysis of the results.

Experience shows that minor adjustments to design work will always be necessary as a direct result of research. Three weeks are allowed for these and the finalization of the copy on all the faces of the pack. Any photography or creation of illustrations is scheduled to take place at this time. It is usual for design studios to produce their own experimental photography or illustrations early on in the study to demonstrate how the pack will look. At this point, outside specialists are commissioned to produce the required material under the art direction of the studio. Specifications for the packaging are finalized, and accurate costings are obtained.

The design phase is then succeeded by the fourth stage, which is a production phase, during which artwork, films and proofing are prepared. When necessary, blank, unprinted packs are obtained for filling and production trials. Any modifications to materials specifications have to be made now, before the packaging production moves into full swing. As the packaging becomes available, the pack filling can start, allowing a period for the buildup and distribution of stock before launch.

Two further activities are carried out in parallel with these. The advertising agency will have been formulating its campaign, and have been in contact with the client and the design company. It may have requested mockup packs for advertising purposes, and these will then have been specially produced as a separate substage, probably soon after the final design is approved. Additionally, secondary packaging will have been developed, and in this study this consists of shipping outers and display trays. The plan shows this taking place following the approval of the final unit pack designs.

The completed project plan indicates a total time for the project of around 22 weeks, which is in line with the original estimate. There is little scope for contingencies, and the plan is couched in terms of working weeks, rather than specific dates. When actual dates are added to the plan, there could be some slippage due to holiday periods, plant shutdowns etc.

This plan is accepted by the marketing group, with some minor modifications. What then actually happened is described below.

The store survey reveals that, as expected, competing packs have the same carton format and size, and are displayed by brand rather than variety. A display tray is used to collate the packs, with some brands showing the brand on the lip of the tray. Significantly for the graphic design, this lip extends upwards for about 30 mm, hiding any text or graphics on the bottom front of the pack.

There is little point in the factory visit as the factory is geared to producing liquid product and not powdered soup. The premises are also restricted in terms of space, and it is doubtful whether it can accommodate any additional lines. The situation in the sister company is more encouraging, however, as it already produces powdered custard in sachets. However, the existing machinery has no further capacity, and a purpose-built production line would be required. It is agreed that this issue should be the subject of a separate investigation, and that contract packing is the only realistic option for the launch of the new product.

Discussions with the advertising agency are encouraging. It is keen to see some new packaging format which makes the product distinctive. With a valuable account at stake, it is anxious to gain every possible advertising advantage, and it looks for any way in which the pack can have added consumer benefits.

The design studio spends the first week looking at potential solutions. Although slides from the store survey are not available, several sales outlets have been visited to determine the point-of-sale conditions. The amount of product per serving seems insignificant when seen in a bottle or tube, and the size impression would quite obviously be poor in this type of format. The concept of some type of 'capsule' to contain the product is thought to be a nonstarter for similar reasons. A favoured alternative is that of 'stick' sachets. These are not new: they originated in Japan. Their use has largely been confined to coffee and sugar individual-portion packs. The incorporation of a tear strip into the tall, slim sachets builds in a consumer advantage. The packs, or sticks, are easy to open, and, being thin, are easy to empty. 'Normal' sachets are rectangular, and, if a corner is snipped off for dispensing, there is often the problem of powder accumulating in the corners. The prospect of a clean one-shot dispensing system is particularly attractive for a calorie-controlled product, as the amount of calories claimed on the pack may form part of a dietary régime, with implications for an accurate dosage of product. Stick sachets are also taller and slimmer than their conventional counterparts, which is an analogy with the claimed benefits of the product, and this fact is not lost on the advertising team. Because they are taller and thinner, multiple stick sachets also have a different appearance on the shelf when cartoned. The pack dimensions can be noticeably taller and slimmer than those of competing products in 'standard' cartons. While the 3D designers on this project become excited at this prospect, the graphic designers are less enthusiastic about a reduced pack width. This affects the area available for branding, and, as an existing logo must be used, there is little scope for altering it to suit the proposed format. A tall slim pack presents graphics challenges in this respect. Experiments are conducted using the revised dimensions, however, and, although the logo is smaller than the ideal, the pack does appear to work.

Initial investigations into the availability of this pack style at UK-based contract packers reveal that two companies are equipped with the machinery to produce and fill stick sachets. Machinery is available from Japan if in-house packing is still a subsequent requirement, but in this case there would be proposed capital expenditure to assess and approve, with consequent delays in delivery, commissioning etc.

The presentation concluding the first stage is on schedule, and it covers all the aspects of the work to date. The stick-sachet route is recommended, and it finds particular favour with the advertising agency. The marketing staff are also interested, but they have reservations about the subsequent adaptability to in-house production. They feel that the principle is right, but they want to discuss the matter further with their sister company's production department. Concern is also expressed about the effect of the pack on shelf stacking. As the store survey confirms, two trays of product are often stacked one on top of the other, leaving little headroom. Will the taller packs allow this? The smaller logo is also a worry, possibly playing down the brand identity that is of key importance to the marketing strategy. The following course of action is agreed:

- Consult the production department of the sister company about the possibility of manufacturing stick sachets.
- Confirm with the purchasing department the availability and selection of potential sources of stick sachets.
- Put on hold any further 3D development until the above point is resolved.
- Continue graphics development, including of extensions to varieties. Explore methods of increasing the logo standout.
- Investigate the proposed format with regard to the shelf-stacking requirements.

At this point, the planning schedule is still on target, and apart from a few differences in the planned activities, the second-stage timings do not require modification.

The second stage sees the abandonment of the stick-sachet concept. The client feels, after discussion with production staff, that a new production line is not justified. As the idea of bringing the production in-house at some stage is important, there is also little point in pursuing contract-packing options for stick sachets and then later making a change in the pack style. The concept of a taller, slimmer sachet is not completely overlooked, however. The sister company has experimented with sachet sizes, and can produce a sachet which, although it lacks the elegance of the pack solution, is at least taller and slimmer than those of the competitors. From the point of view of shelf stacking, this configuration is just feasible, and it has the benefit of reducing the facing-panel widths, allowing more varieties to be displayed in a given length of shelf. The logo is now larger, and the standout is now felt to be better. Two graphics routes are considered, one using product photography, and one (preferred by the design studio) using illustration. Illustration is claimed by the studio to be more effective for this product, as the product itself lacks appetite appeal when photographed. It is agreed that both routes will be explored during market research. As the new pack will have proportions that are different from those of competitor packs, research is conducted using full-scale printed mockups.

Now there begins to be some deviation from the original plan. To produce fully printed mockups, it is necessary to produce artwork that consists of typeset copy and stripped-in photographs and illustrations. First, accurate copy is as yet unavailable, and, second, the illustrations and photography prepared thus far are experimental and not of the standard or accuracy required. The illustrations, it is agreed, will be prepared by the design consultancy. (The loose style of illustration shown in earlier work is liked). The photography has to be commissioned outside, rather earlier than scheduled. A specialized food photographer and a home economist to prepare the product are required. This cannot be accomplished within the planned timescale.

Looking ahead, however, it can be foreseen that time can be saved on the production phase of the project. Time has been allocated for artwork production which, in effect, has merely been moved forward. As little in the way of technical pack development is now required, time can also be saved here, and the production trials can be minimized.

In this project, the plan is flexible enough to accommodate minor changes that cannot be foreseen at the outset. Because there was a plan, it was possible to predict the effect of change on the overall targets.

It might be thought that, in projects that are more complex and take longer, there are more time reserves to cope with contingencies. The next example, which has been slightly simplified, shows that this is not the case. The more complex the project, the more chance there is that things can go wrong.

COMPLEX STUDY

In this case study, the client is an international oil company that wishes to redesign its range of lubricants. Its sales outlets are confined to the company's garage forecourts throughout Europe. Although no competing products are ever displayed next to the client's own products, it is felt that the current pack design is dated, and is out of line with those of other products marketed by rival petroleum companies on forecourts and in specialist shops. The client is looking for structural and graphic designs that convey modernity and will remain up-to-date for the next ten years or so. Packs will be produced and filled within each European market sector, involving multiple suppliers and a variety of filling plants. Timing is important, because a simultaneous launch throughout Europe is required. No date for this launch has been imposed. The pack sizes will be 0.5 l, 1 l, 2 l, 4 l and 5 l. The languages on the pack will be those of the country of sale, or there will be multilingual versions for groups of countries.

A project of this nature is complex, and, despite the lack of imposed deadlines, careful attention to planning is required if all the container sizes are to be launched in all of the participating countries at the same time. Before the project start, the outline is as follows:

- *Imposed schedules:* There are none, other than the fact that an international launch of the new pack in a range of sizes must be coordinated.
- *Design work:* The technical, 3D investigation must precede the graphics study. It is expected that the 3D study will concentrate on one pack size initially, and then this will be extended across the range. The timescale estimate is around four weeks for the first-stage concepts, with a further four weeks for range extensions. The graphics study could start during the latter part of the 3D study, and extend for a further three weeks after firm 3D directions have been established.
- *Possible outcome:* The pack format is most likely to be a polyethylene container.
- *Market research:* This is required in order to test the 3D designs and graphics in the major European countries of sale. Six weeks is allowed for this activity.

- *Production time:* The production of polyethylene containers is unlikely to take much under 16 weeks. The stock build-up of containers will take a further four weeks, and filling and the distribution of stock is estimated at four weeks.

While it is not possible to be certain that a polyethylene container will be the preferred design route, it is the most likely option. If it is the case, it can be seen from the rough estimate above that the timescales for the activities already listed produce a timing estimate of 41 weeks. As the international element is likely to cause administration problems and require additional communications, it may well not be possible to complete the overall project in less than a year.

Discussions with a design consultancy yield the timing plan shown in Figure 4.2.

The client company is well aware of the complexities involved within their own organization in terms of coordinating design activities effectively across national frontiers. Designs that are acceptable in the UK might find less favour in France, for example. Gaining design approval for each country represents a difficult task in itself. For this reason, a project director is appointed with a mandate to coordinate the project internationally. A panel of representatives drawn from each participating country will convene at key stages to approve or otherwise comment on the design work as it progresses. Production specialists will also sit in on these meetings to advise on the complex production situations in the participating countries. Supplier discussions will similarly be eased by the use of a purchasing agent to assist in the coordination of the multisourcing and multinational-sourcing aspects of the project.

Administrative problems of the type outlined above should never be underestimated. While many design companies work under pressure to meet tight deadlines, time can be lost during the design stages and afterwards in chasing around trying to obtain approvals and copy, or in attending meetings. It is in this type of situation that a project plan becomes an essential tool in keeping track of progress and keeping progress on

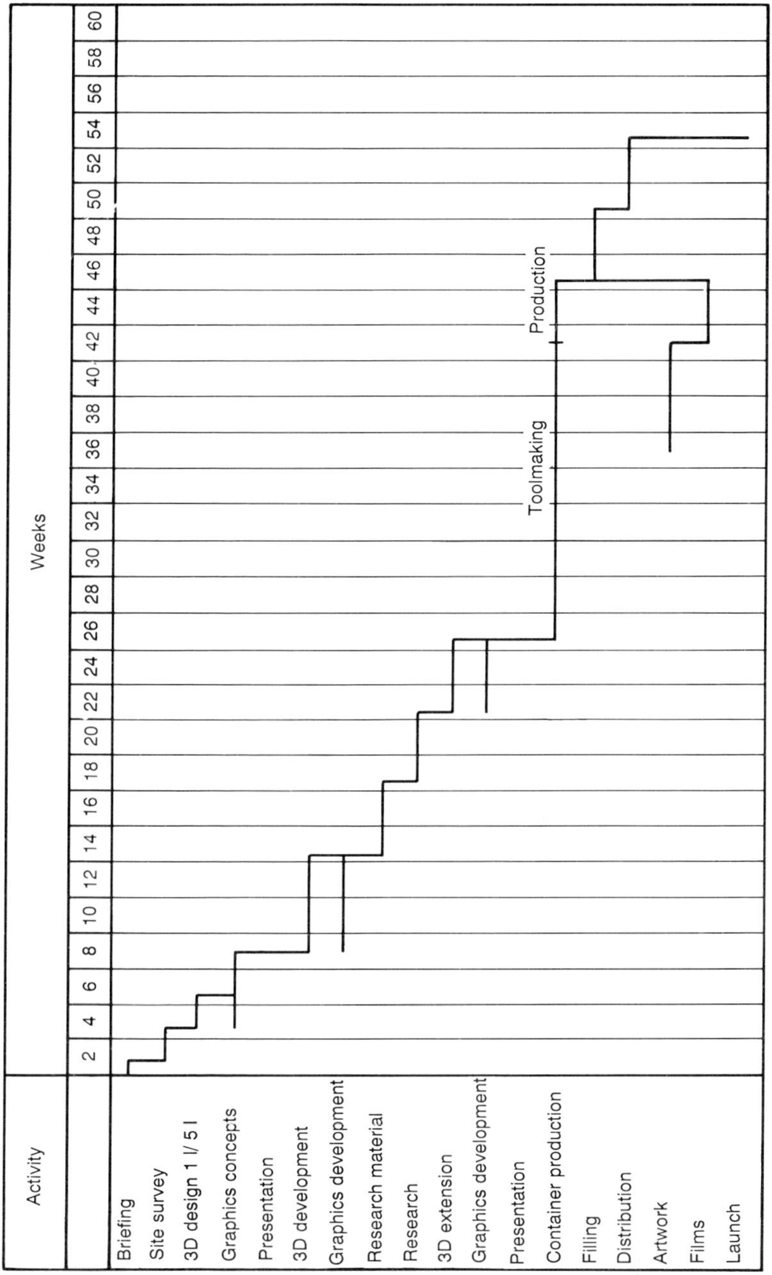

Figure 4.2 Initial project plan for complex case study

track. When many individuals are involved in the project, each person needs to be kept up-to-date with events. A copy of the project plan and each amendment made to it becomes an effective method of communication.

The importance of the role of the project coordinator or director should also not be underestimated. In the type of project in this example, the coordination activities will place a heavy demand on a person's time. It is unrealistic to expect an individual to take on this role in addition to carrying out his or her normal day-to-day function. However, it is also understandable if client companies cannot afford to employ someone just to coordinate a single short-term project. The alternative of using an outsider, for example one of the design consultancy's staff, often does not work either, as he or she will not have a detailed knowledge of the client company, its way of conducting business, and the individuals involved. In this project, the coordination role is a full-time assignment for one person, with a part-time backup.

As mentioned above, the project plan is a dynamic tool, subject to change, which is designed to help in the organization of the project. In this example, many factors do change during the course of the project, and they are typical of the problems that project planning has to cope with. For this reason, they are dealt with individually below. They are areas which require consideration when project plans are drawn up or subsequently modified.

Design by committee

Most design projects involve a team of designers who each have individual approaches to a design problem. This is to be encouraged, as it enriches the study, providing a depth and breadth that might otherwise be lacking. While it is normal for the design team to show the client the full range of their work, the team will already have narrowed down the design routes into those that they feel are worthy of promotion, and those that they feel should fall by the wayside. The logic behind their selection is always explained to the client so that design thinking can be followed. There is seldom just one design solution to any problem, and so it is common to have, at the end of the first stage, two, three or more routes which are worthy of further consideration. There will often be more than one person from the client company at the presentation, and they can be expected to have more than one preference. Sometimes the features of two or more designs are combined to form a hybrid which satisfies everybody. In the study in this example, a panel of client representatives is drawn from the various relevant countries, disciplines and backgrounds. Many opinions can thus be expressed, with the consequent danger that it will not be possible to agree on clear directions for further work. All the opinions are valid and are based on solid reasoning. As noted above, a design that suits Italian customers might not be right for Swedish customers, and national differences must be accounted for. Many believe that it would be unthinkable for a Ferrari to emanate from a German design house, just as it would be amazing if a Porsche emerged from Italy. They both have excellent designs, but they owe their design origins to national characteristics and national schools of design.

In this example, four design routes are selected for development — more than the plan expected. This entails the design origination of four complete families of containers (20 containers in all), which creates the next problem.

Insufficient design time

It is almost impossible to predict the outcome of a design study. At best, the guess can only be based on experience of similar projects. In this example, the design studio expects two 3D design routes, and is faced with the development of four. As the studio has been instrumental in developing the project plan, it is difficult for it to deviate from it and extend the second-stage investigation. Pressure from the client to adhere closely to the agreed timing means long hours and unbudgeted hours, which is a potential billing problem. To minimize the development time, the design studio concentrate on the 5 l and 1 l sizes, producing rough models of each for the four selected design routes. Sketches are produced to illustrate the remaining sizes that depict the complete family range.

Similarly, with the graphic designs, the committee fail to agree on one solution, and require further work on five design routes, again raising the dual problem of timings and costs. Before each stage is started, it is usual to quote for the stage in advance. While the reason for increased cost (i.e. increased time spent) can be understood and agreed, the additional costs often start to eat into the overall budget allocated for the whole project, creating a potential timebomb for the accounts departments at a later stage.

Presentation timing

The purpose of presentations is to provide an opportunity for the client and the designers to get together to review the work, decide which directions to pursue, and agree the next phase of the operation. When many people are involved, the logistics of arranging such meetings become difficult. The problems increase when, as in this case study, members of the panel reviewing the design work have to travel from distant destinations, and have busy personal schedules organized outside the parameters of the project. This can result in delays and occasions when someone is unable to attend a meeting where decisions are taken which may directly affect his or her operation. Whenever possible, client teams should be kept to a few people who have enough authority to make decisions on behalf of the company.

Market research

As mentioned above, in this project, four 3D-design routes and five graphics routes are worked up to the second stage. To assist in the selection of solutions, market research is considered to be essential, as indeed it has been from the start. There are two main criteria that must be tested: the selection of the

3D designs, in comparison with the current design, for a family of containers, and the selection of the graphics design. Because of the pan-European dimension, research in more than one country is necessary. With 3D aspects being of prime importance to the success of the project, it is necessary to produce models of each of the four design directions, and, to make the comparison fair, models of the existing containers as well. Each model has to be suitable for handling, as this is a key issue in consumer acceptance. The models, therefore, have to be representative of the filled containers, weighing the same, and having the same centre of gravity. Any features that are important to the design, such as pull-up spouts, also have to work. In this project, to meet these conditions, it is necessary to produce a set of models for each design, including the current design (25 models in all). To enable research to be carried out within the same timescale at each of four selected locations, 100 models would be required. Model making to this standard is time-consuming and expensive, and it would push the timing plan and the budget out of line. While there are several possible solutions, in this example, the model making is confined to two sizes, with photographs, using one full set of models to depict the full container range. Rather than four sets of research material being produced, only two are commissioned, and these are moved from location to location in sequence. This activity starts to stretch the timing plan, which requires continual updating and detailed day-to-day scheduling to minimize delays. With model makers, designers, market-research company staff and photographers working all hours, incurring 'rush-rate' costs, the budget is seriously eroded. Even delivery by private aircraft is undertaken to ensure that models reach the researchers on time.

The graphics research material shares a similar fate, requiring artwork and printing to be undertaken for relatively few labels. Achieving the required type of finish, which includes pearlescent and metallic pigments, is a challenge.

After these problems have been surmounted, the research does not identify a clear winning design. One design is preferred in France, and another in the UK; other countries have their own preferences. Two of the four designs are eliminated, but the client company is left with a dilemma to resolve. This is done by carrying out further design modifications and yet another, originally unscheduled, stage of market research.

This result should not be read as a catalogue of errors. It is simply how some projects develop when unforseen events occur. Fortunately, life is never dull in packaging design. The study illustrates the need for planning, and, in particular, the need to update and reschedule plans constantly as events unfold.

Container production

After one design route has been agreed, accurate drawings are needed, in this case for blow-moulded bottle suppliers. While design consultancies are able to produce drawings that are suitable for model-making purposes, they cannot incorporate all the necessary detail for toolmaking, because design teams are not familiar with the technical requirements of the machines used for container production, and they cannot calculate shrinkage requirements. (The tooling has to be made oversize to allow the bottle, which is hot when produced, to cool to the correct size.) The containers must also hold the correct amount of product together with a preagreed headspace (ullage), and this is a difficult calculation for a complex nonregular container shape. Designers therefore have to rely on suppliers to provide this expertise, together with a knowledge of draw angles and other details which enable the containers to be ejected from the moulds. When there are multiple sources, each supplier needs copies of the designers' concepts to 'translate' into actual tool drawings that are suitable for their machinery. The overall design concept must not be lost, and bottles from one supplier must, to all intents and purposes, be identical to those of another.

This process is time-consuming and expensive, and it is further complicated by the fact that some suppliers are capable of producing smaller containers but not the larger sizes. The economics of qualities and costs must be considered, as must locations and manufacturing capacities. The usual leadtime of 16 weeks, say, is affected by such issues, which adversely affect timing. Rescheduling the project plan becomes a major and vital exercise.

Artwork and label production

The project plan shows that the artwork preparation is scheduled to take place fairly late in the project's progress. Indeed, it can only happen when the graphic designs have been agreed. While it is possible to position the main elements of the design, copy in the language of the recipient countries is required to complete the artwork. The failure of marketing groups to provide accurate copy is probably the most frequent reason for projects being overbudget and delayed. When there is foreign-language copy, the problems increase. The copy is sometimes set by a person who is not familiar with the language, and sets it character by character. It is sometimes set by a company in the country of the language, and supplied as complete artwork to be stripped into the design. It is sometimes sent outside to specialists in the language concerned. Rarely, however, is it set correctly the first time. The copy must be set and then checked by the originator, and preferably signed off as approved. In practice, this is time-consuming, entailing copies being faxed or posted to everyone involved. There is often a temptation or a need to modify the copy, resulting in yet further delays. The project plan must again be updated.

Labels may, as in this example, be produced at various printers, perhaps using different processes. In this project, offset litho, letterpress, gravure and flexographic print methods were used, and yet the results had to be similar. Some printers wanted the artwork to be supplied on disc, and others on film. All of them had to submit proofs for approval. The timescales were telescoped, and mistakes crept into the artwork as well as the label production.

The point of the above is not to highlight mistakes that have been made, but to alert those responsible for planning a project to the problems that can arise. Copy should be prepared early in the project, signed off, and be available long before the labels are required.

Budgets

The above has demonstrated that time is equated with cost.

To maintain the time schedule, it may be necessary to telescope time by the studio and other services working overtime. This almost always incurs 'rush-rate' costs, particularly in the case of outside agencies such as photographers, developing laboratories and repro houses. Costs can quickly escalate. The design-consultancy costs will also be increasing. Designers almost never feel that they have arrived at the definitive solution, and, given more time, they will beaver away trying to achieve perfection. In fact, most design consultancies work rather better under a certain amount of pressure. Getting this degree of pressure right is difficult. If there is too little, the project limps along, and, if there is too much, there is not enough time to explore all the possibilities. However, experience shows that most consultancies delay projects until the deadline approaches, and then engage in a concerted effort to solve the problems. It is then that there are strokes of genius, and many winning designs have been presented to clients with the ink not dry on the paper.

This chapter has shown how to approach project planning, and discussed some of the most common pitfalls. There are many aids to producing a plan. It can simply be prepared on graph paper, or, which is better by far, it can be prepared using one of the many software packages available. Whatever the method, planning is important in maintaining an effective packaging-development programme. Equally important is the ability to alter the plan to cope with changing conditions. No plan should be regarded as final. Flexibility is the key to control.

5
Predesign research

It sometimes surprises those not used to the ways of design companies when considerable time is allocated in the project plan where no actual design work is scheduled to take place. This time is spent in predesign research, and in getting to know the project, its environment, the competing products, the point-of-sale conditions, the manufacturing criteria, and the opinions of those who are expert within the areas which impinge upon the packaging. While the designers are not putting pen to paper at this point, they are beginning to grasp the problems and formulate solutions. While the theory and the timing plan may state that no design work will commence until the research has been completed, designers can rarely resist putting their ideas down on paper in practice, even if these ideas have to be revised or even abandoned in the light of subsequent knowledge.

STORE SURVEYS

As the majority of packaging-design work takes place in the FMCG market sector, the term 'store survey' has become the accepted terminology for looking at the point-of-sale conditions within retail outlets (usually supermarkets). Not all products are sold in this way, and a 'store survey' should really be considered in a broader context as a survey of any area which influences sales. For car accessories, for example, the 'stores' may be dealer showrooms, and, for horticultural products, they may be garden centres.

The purpose of a store survey is to gather background information for the design team. There is often a spin-off benefit, in that the client company commissioning the design study can be updated about the latest activities taking place within the market sector in question. Many surveys have shown the client what is really happening at the point of sale; this may not be the same as what the client thought was happening.

The approach to the survey must be methodical. While there is some merit in wandering around stores looking at product displays and packaging, it is difficult to be objective while doing this. The researcher is distracted by the variety and mass display of products. For a truer perspective, it is better to take colour photographs. Colour slides viewed in the studio allow aspects of packaging to be studied in much greater detail than can be done on the spot by eye. It is also a cost-effective way of allowing the whole design team to access information about stores. It is not viable in terms of cost to take a complete team around a number of stores in different locations and regions, or different countries for international studies.

Stores are surveyed to obtain information about the following (see Figure 5.1):

- *positioning:* where the product is most likely to be displayed,
- *proximity:* what products will be displayed above, below and on each side of the subject pack,
- *competitors:* how competing products and brands are packed,
- *lighting levels:* what are the light and dark areas of the display area.

If the product falls into a new category, it may not be certain what its in-store position will be, but the client's marketing department will usually have a clear idea about its expected positioning. The positions of packs of products which fall into existing product categories can be predicted with accuracy.

'Positioning' refers both to the location of the display area within the sales outlet, and the position of the product on the

Store survey

Canned ready meals

Lighting level

1/30th sec: 200 ASA					

Brand	Facings	Brand	Facings

Proximity test

	Brand	

General observations

Outlet name . . . .

Outlet address . . . .

Classification

Grocers
Multiple/Coop/Independent Symbol/Independent
Self-service/Counter Service

Figure 5.1 Store-survey research form

shelf with reference to the consumer's eye level. If the product is a frozen precooked meal, for example, it will be in the freezer cabinet designated for this type of product. This could be a chest freezer, where the pack is laid flat and examined by the purchaser from above, or a vertical-cabinet freezer, in which the pack stands on a side panel at, or just below, eye level. The pack may have to suit both types of display condition, and it will be important to ensure that the main display face of the pack works in both circumstances. Pack configuration and panel orientation must be considered carefully. There are many instances in which design work is commissioned without enough attention being paid to the actual display conditions, and the major design effort goes into designing panels which are then not seen at the point of sale. Classic examples are bags of flour and sugar. These products are often stacked horizontally, with only the end faces being displayed, and sometimes no branding being visible at all. This also happens with cartons of tea, with end panels being displayed that show a bar code, but no branding. While it is unrealistic to expect the stores to change their display methods, if these problems can be identified before design work starts, the designers can make allowances for the display methods used. There are two approaches that designers can consider. They can accept that the end panels are the panels that are going to be seen most frequently, and ensure via graphics that the end panel at least carries branding, or is in some other way distinctive, perhaps because of the use of colour.

Alternatively, they can consider redesigning the pack format to ensure that the main panel will be the one on display. In one example, in which the product to be sold was flour, a switch was made to a rigid, composite pack. This allowed the pack to stand, and presented a large front panel for graphics and branding. The improvement in standout immediately increased sales, to the detriment of the sales of the flour that was still packed in bags. No doubt the new pack form was more expensive, but it also offered the opportunity to incorporate consumer benefits, easier storage, and improved access via a hinged lid, which helped to sell the concept and justify the cost.

The way in which supermarkets are laid out leaves little to chance. Extensive work has been done to determine consumer behaviour, and to design the store to make the best use of consumers' behavioural patterns to sell products. The theory used by many of the sophisticated retail groceries is that the store should have the entrance on the left-hand side of the building, as people tend to look left and merely scan to the right. Fresh vegetables are placed near the entrance to set the tone of freshness encouraged by the stores. Most shoppers are then led to the back of the store, preferably to where the 'bakery' is sited, wafting the alluring aroma of freshly baked bread into the main store area. Commodity goods tend to be at the back of the store. Aisle width is important: narrow isles promote faster transit, and wider aisles encourage browsing. The goods with higher profit margins are usually located in the wider aisles. Products are classified vertically, with higher-margin goods being placed at eye level, or slightly above, and the less profitable lines being relegated to the lower shelves. With such sophisticated merchandising concepts being used, it is understandable that the product manufacturer has little control over product positioning in-store. A store survey reveals positioning for the design team to take into account. When packs are going to be on bottom shelves, it may be important to include branding on the top. In one example, dog foods were shown by a survey to be often stacked on bottom shelves. From the consumer's eye level, it was mainly can tops that were seen. The tops of the cans were printed, and additional branding was achieved and sales increased.

Multiple groceries are no less 'scientific' in their approach in relation to the profit yield per area of product. The number of facings that a product is allocated depends on its profit yield. Brand leaders are in a stronger position in this respect, as stores are virtually obliged to stock them to maintain customer acceptance; if they do not, customers may shop elsewhere. With a product that is not in this fortunate position, the store has to have some profit incentive to stock it, and it critically assesses its pack dimensions in order to obtain the maximum return from the area that it occupies. When positioning is considered, this is another important aspect. Pack dimensions should allow the most efficient shelf utilization. When it is planned to introduce a range of product varieties, the design team must examine the store-survey slides closely to see what opportunities there are for stacking products two layers high within the existing shelf-space limitations for example. At the store-survey stage, only the possibilities are explored. If there are potential opportunities for maximizing shelf utilization, actual measurements will have to be taken later. If too little attention is paid to this aspect, it may not be possible for the manufacturer to sell in the total range of product variants.

The proximity of other products and brands in relationship to the product studied is an important design consideration. These are the packs that the product will have to compete against. If it is to succeed, it will have to stand out from its competitors. Other factors (price, quality and advertising) are unlikely to be influenced by the pack, and so these areas must be covered by the manufacturer, leaving the design team to do their best with the pack. A poor product will fail, no matter how brilliant the packaging design is. When packaging overpromises, this is a major disappointment to consumers. Often, the competitor packs above, below and on each side of the pack on the shelf are in the same product category as the product in the pack. In this case, competition is head-on, with the consumer comparing like for like. This is not always the case. Instant-cappuccino powders, for example, may be displayed among cocoa powders, coffees, teas or other products. If a pack is positioned at the end of a section, it may have unrelated products adjacent to it with which it must compete. Goods in proximity to the pack always represent competition, even when they fall into a different product category, in the sense that they compete for the shoppers' attention. The first task of any pack is to stand out. The survey slides will show what the pack has to stand out against. The design team will have to use these as one way of measuring the strength of any design solutions.

The products that are likely to be in proximity to the pack will be revealed by the store survey, but actual samples will need to be provided or purchased for closer examination by the designers. On most occasions, it is mass displays that the shopper sees, and not just individual packs. A good design team will use the store-survey slides to recreate mass displays, inserting any new designs into the mass-display format. A client should be suspicious of any design source that works using individual packs alone, whether in analysing competitors or producing new designs. Good pack design can use the effect of mass display to amplify design characteristics. Poor pack design diminishes standout when the pack is seen in a mass display.

Lighting levels must be considered when packaging is designed. Some areas of supermarkets used to be poorly lit. This is less likely now, but the market for the product may not consist only of supermarkets. Cash-and-carry sales may be a significant contributor to the total sales and the lighting may not be so good in these outlets. There are also low-price self-service outlets where the lighting is deliberately restricted to reduce costs, and also, one suspects, to encourage purchasers to think that no money is being wasted so that the products can be sold at the best prices. Lighting levels can be measured. Cameras with a built-in light meter and a standard 200 ASA colour-transparency film and a constant shutter speed can effectively measure lighting levels in terms of aperture requirements from f3.5 to f22. Any measurements below f8 indicate a poor light level, and f8 and above is acceptable.

In-store lighting levels are a factor which the pack designer cannot influence, but must take into account. There are many reasons for using darker shades of colour on a pack; they are often associated with quality, or richness. If, however, the lighting conditions at the point of sale are poor, the imagery is wasted if the brand logo becomes recessive. As standout is a prime mover in packaging design, the pack must function well in lighting conditions that are representative of those found in actual practice. Design consultancies may present packs to clients on shelves at the consultancy's premises that are lit by spotlights. This may show off the pack design to the best advantage, but the actual display conditions will usually be different. Lighting levels used during such presentations should be approximately the same as those found in the relevant stores. Strip lighting, or even natural light, may have to be used for demonstrations if those are the conditions at the point of sale.

The survey must be conducted in a selection of outlets that are typical of the point-of-sale conditions under which the product will be sold. If the pack is going to be displayed within the multiple-grocery type of outlet, for example, a representative spread of such outlets must be visited. This might comprise out-of-town superstores and high-street supermarkets, and include at least two stores owned by each major grocery company. Regional variations, and national and international dimensions if these are important aspects of the product's market, must be considered. A typical spread for a UK store survey for an FMCG food product would include two or three stores from each of the Waitrose, Sainsbury, Tesco and Asda chains, and smaller stores if this is important to the study. In some cases, outlets where the product will not be sold (for example Marks & Spencer) might be visited to see how they package similar products.

Usually, the design consultancy running the study will send a researcher to each location. It used to be possible to obtain permission from each retailing company's head office to take in-store photographs. Permission is less often forthcoming nowadays, and trying to obtain it is a time-consuming business. In the UK and many other European countries, there is a marked reluctance to allow photography within stores, although it is difficult to see any concrete reasoning behind this. Photography is therefore usually a covert affair, with the researcher occasionally being shown the door, politely or otherwise. When official permission has been granted, the store in question is alerted by head office. This often results in an immaculate presentation of all the goods in the store on the day of the survey. The researcher is greeted by a manager, and then shown around shelves which bear little resemblance to the average state of the displays. Such arrangements are not useful for research purposes. Figure 5.1 shows a typical format for recording store-survey information.

RETAILER DISCUSSIONS

The store survey is a snapshot of the retail scene at a particular moment. There can be no guarantee that what has been recorded on film will not change, even within the timeframe of a packaging-design study. To find out whether there are likely to be changes which may affect the design options, it is necessary for the designers to talk to the retailer.

This is normally only done for a major repackaging exercise, or for a leading brand. Representatives of two disciplines within retailing must be consulted, the product buyer and the merchandiser.

Buyers are used to being sold products, and it can be a refreshing change for them for their opinions to be sought. The job of buying is demanding, and it carries significant responsibility in terms of company performance. Buyers have considerable power, and they have strong opinions about every aspect of packaging. Access to a buyer's office can only be obtained with the active assistance of the client company, as buyers do not have the time or inclination to spend time discussing packaging matters with design groups unless the buyer can relate this to issues that directly affect his or her own operation. In such a discussion, it helps if the researcher outlines the project, and perhaps even the early design routes which the design team think may have potential. Points will arise that are useful additions to the study. Others will be less useful, for example because they represent the individual quirks of the buyer rather than the view of the retailing company. The researcher must listen, and then later sift the information for significant data and opinions. From the client's point of view, it is probably not a bad thing to involve buyers who then take an interest in the project. It may assist the client to sell a new or revised product to the store.

Most major retail groups have a merchandising section. This makes the decisions about product positioning. A visit to such a section is recommended as, apart from having the chance to obtain specific information, the researcher will find the operation fascinating. There is often a test setup that is virtually a complete store, equipped with shelving units, gondola ends, freezer cabinets etc., and with products stacked on the shelves. Any new display systems are tested here before the stores are fitted out with the new units. Details of current and planned shelf dimensions can be obtained, usually with planograms showing exactly how the shelves are to be stacked. Planograms are computer-generated layouts of the shelves with packs superimposed, and they can take the form of colour scenes or black and white line drawings that are marked up with details of the product positioning. These are issued to all stores as instructions to be followed by store managers. Decision making about product location is firmly under the central control of the merchandising department, and it is not left to the whims of individual managers. It seems that merchandising staff are seldom asked their opinions prior to the generation of new packaging, and they are therefore receptive to being approached for their opinions.

The project plan must allow sufficient time for retailer discussions if these are to be an input into the study. Making appointments with buyers is not easy, and, if several companies are involved, the process may spread over some weeks.

ADVERTISING-AGENCY DISCUSSIONS

It is interesting to see how often the packaging of products features in advertisements on television and in the print media. This is not surprising, because the packaging acts as a silent salesperson in all self-service outlets. Advertising makes consumers aware of a product, extols its virtues, and induces a desire to buy, but it is the packaging that makes the sale. Packaging design and advertising are complementary activities that form a powerful alliance. Both disciplines are creative, and both have to be structured to meet the same brief. Designers and advertising staff thus share the purpose of creating something which will sell product. The languages of communication of the two parties are much the same, and so they immediately have a rapport, even when they are questioning each other's conceptual thinking. In the UK, design and advertising are usually separate activities. This is not the case in many other countries. In continental Europe, it is often the advertising agency that is responsible for packaging design, and one can see the logic for this. There is also an argument for the two disciplines being separate, however, as packaging design does require the use of different skills, some of which are creative, some technical and some analytical.

Interestingly, advertising staff often have a perspective on packaging that is different from that of pack designers, and any study is enriched by accessing this. Advertising campaigns, particularly on television, are fun, and so packaging designers readily seize the opportunity to participate in these. Clients may not think that the spending of considerable sums of money should entail having fun. However, original thought and serious propositions that utilize the full potential of packaging as a sales and marketing tool arise from this process.

Probably the most important contribution that advertising staff make to a packaging study is their background knowledge of the product and the consumer. This is usually impressive in times of its depth, and it is expressed in such a way that designers understand it. Designers think in pictures, and so do advertising staff. Their thought processes are thus entirely compatible.

PRODUCTION SITE VISIT

If the designers do not have a knowledge of what packages the client company has the ability to produce, the development of any radical new packaging concept is largely an academic exercise. If the design work is going to entail changes in the pack structure, it is essential for the designers to visit the production location.

One of the principal benefits of site visits is that they allow the designers to meet the people responsible for producing and packing the product. While the primary purpose of the visit may be to investigate the machinery and the lines, obtaining the production manager's backing for the project provides a further, expert input to the study. In terms of human-resources management, it ensures that everyone is playing for the same team. For example, it would be entirely understandable for a production manager in the provinces to feel resentful about packaging-design solutions forced upon him or her by some 'hot-shot' design house in London. Production management is often regarded as being awkward, reluctant to accept change, and quick to find fault with new ideas. However, the majority of production managers are open to new thinking, provided that they are included in the development process, and given the opportunity to express their views. Production departments also carry the responsibility of getting product out of the door, and they have targets to meet and budgets to work within. Given the age of some production and packaging machinery, and the lack of funding at some production locations, it is often surprising that they can manage as well as they do. They can be forgiven the odd complaint as yet another consultant visits their factory.

The site visit should be comprehensive, and should cover every area from goods inwards to goods leaving the factory. The goods-inwards area and its associated storage facilities give an indication of the space available and the storage conditions. If the current packaging is based on folding cartons, a change to rigid cartons, for example, could mean that the storage area would have to be expanded, or it could be not a viable alternative if space is already at a premium.

It is the filling and packaging lines that are of major importance in establishing what is possible with the use of existing equipment, and what the opportunities are for modifying the existing facilities or installing new plant.

The specific production issues depend on the brief. One brief prepared by an oil company called for the replacement of metal cans by plastics containers which had to run down the same production lines. Here, the 'footprint' size of the pack was critical, requiring careful checking at each stage of production.

Depending on the nature of the project, therefore, the designers will look at specific production details. Generally, the manufacturing capabilities of the existing filling and packing lines will be assessed to ensure that the pack designs will meet the criteria of the current production techniques. There are sometimes opportunities for totally new machinery to be introduced, but this is rare. More often, ancillary online or offline equipment must be considered, such as overprinting or coding machinery.

Secondary packaging is also an important issue. Many studies focus attention on the primary pack, with any form of secondary packaging being forgotten. Changes in primary packaging may have repercussions for secondary packaging. Changing the size of the primary pack, for example, may mean that a wrap-around cartoner machine may not be able to handle collations of the new pack size.

In some organizations, warehousing may be important when considering the pack configuration. The ability to maximize pallet loading can, for some product types, be a critical cost element. This is also the case for transport methods; the number of packs per vehicle can affect profitability, for example. Indeed, there are companies that start designing the pack from the transportation angle, analysing the optimum vehicle utilization, working back through intermediate containers, and ending up with a recommended primary pack size. This approach, however, pays insufficient attention to the consumer and marketing requirements, which are, after all, the driving force behind sales. There is room for compromise in this respect; the pack-design candidates developed can be assessed in terms of their impact on storage and transportation.

Many warehouses are equipped with automated storage-and-retrieval systems, for which the pack dimensions are critical. Pallet overhang would not be permitted in these systems.

As with store surveys, it is often a good policy to use photography of the production facilities as an *aide memoire*, and as a means of communicating to the design team the importance of what has been found. A site visit will facilitate a greater understanding of the product and its manufacture, and the packaging and warehousing operation. This will help to ensure that the design work is relevant and on target.

Predesign research is an essential step in responsible packaging design, and, although it may seem to be nonproductive in terms of producing design solutions, it will ultimately pay dividends by ensuring that the recommended design options are feasible. It should not inhibit creativity, however, and each element of the predesign research must be introduced to the design team at the appropriate moment.

6
The design process

The process of design starts with the brief. When design consultancies are used, the client will probably discuss the brief at length with the principal members of the design team. The brief will also be encompassed within a written document, and distributed to all those allocated to the project. The nature of the project and type of work that it will involve will dictate the makeup of the team. The design director will assign designers to the team within which their talents will be of most value. The director will try to create a balanced team that includes males and females, and junior and senior designers. An element of fashion is involved in design, and junior designers can bring a freshness and a contemporary slant to the design process; more senior designers can contribute their experience and technique. The full design team will probably not be present at the initial briefing, and a second, in-studio briefing session will be required, preferably where the problem can be explained using examples of current packaging, competing packaging, and other visual aids. A store survey may not yet have been conducted, and other predesign research will probably not have been carried out. While the study will not get under way until this material is available, it is useful for the design team to have a period of grace during which they can consider the project. For example, if they will shortly be working on a project for a premium granulated coffee, designers will probably spend time (often their own time) looking at the coffee area within stores. They may also look at unrelated areas, such as the premium-chocolates section, to help them to formulate ideas. Often, by designer request, the store survey will include apparently unrelated areas that are of interest to particular designers, and which they feel have some values parallel to those of the product in the new study.

By the time that the store-survey slides are available, some designers will already have started formulating ideas, and may have prepared initial sketches. The team will then be able to see all the survey slides, and start to analyse, as individuals and as a team, their implications. There may also be inputs from discussions with advertising agencies and retailers, and the manufacturing site visit. The design director must take a view on the introduction of this information to the study. Too many constraints applied too early in the study will hinder creativity, while details remaining unrevealed for too long may lead to irrelevant work being done.

Individual designers have their own methods of working. Most start by sketching out ideas in the rough (these drawings are often prolific in terms of content), trying to quickly record and experiment with concepts. When they feel that their roughs have some potential, the designs are worked up, either manually or on screen. For early conceptual work, many feel that the software packages that run on Apple Macintosh and PC-compatible systems are too slow, and not intuitive enough. Graphics packages tend towards the exact, rather than allowing freedom of expression. Nevertheless, there are those who always work with these, and find them a perfectly acceptable medium for design. Other systems, such as the Quantel Paintbox, incorporate an 'electronic pen'. These are closer to traditional working methods, and are quicker than the current generation of Apple Macintosh and PC-compatible software. The electronics revolution moves on at a considerable pace, and both hardware and software will no doubt be developed that will permit the 'loose' approach that designers need to produce fast conceptual design work.

At this point, the designers will largely be working on their own, with the design director looking over their shoulders from time to time. It is important to maintain a team spirit throughout the project. Designers are notorious for protecting their work, and each is keen to produce the winning design. Designers have a reluctance to share their thinking with others, particularly if they are convinced that they are on to a winner. Some design consultancies are active participants in design-award programmes, seeing benefits in the publicity that results. Others positively discourage such activities, believing that designers' minds should be fixed on meeting the brief, and not distracted by the allure of awards and seeing their name in lights. Many awards seem to be granted to 'clever' designs, on the basis of artistic merit rather than hard facts. Consultancies are in the business of producing designs that sell products, and this should be the yardstick by which success is measured.

At a point in time that is fixed by the design director, the members of the team will meet to show their conceptual work and discuss the reasoning behind the design directions they have chosen to promote. There may be concepts which can benefit from crossfertilization between the works of several design-team members. Usually, at this point, it starts to become clear which directions have merit and should be pursued, and which routes should be discarded. It is necessary to review all the work, and start the narrowing-down process. It is important that some team members have sufficient expertise to consider the technical aspects of the proposed design solutions. Frequently, and in particular with 3D design, some solutions proposed by designers are difficult, or even impossible, to manufacture. This does not matter, as some of the desirable features of an 'impossible' design can be incorporated into a design which is producible. In the early stages of a design study, it is important to maximize creativity, even if some of the concepts present practical problems. It is essential, however, for there to be at least one team member who has sufficient technical knowledge to recognize manufacturing problems before the design becomes advanced. This task may fall to a technical or 3D designer, who will probably be the person who is seconded to the factory visit. The production limitations and opportunities can be introduced at this point, with the results of the site visit being used as a background to assess the compatibility of the design with the manufacturing capabilities.

As some design routes emerge, it is necessary to start to evaluate these in greater detail. Computer-based systems come into their own at this point. The most promising designs can be originated electronically, and on-screen comparisons can be made. As there is little substitute for seeing a mockup of the pack, colour proofs of favoured concepts can be produced that are then mounted onto blank packs of the correct configuration. The standard of presentation will then be consistent, and this will allow a fair comparison to be made between designs. When other methods of producing mockups are used, those that are slickly produced may be favoured, because of their finish rather than their content. Even at this early stage, it is worthwhile considering how multiple packs will work, and also, if this is part of the study, how product variants and size ranges will appear. This can easily be assessed on-screen. Packs can be replicated and pasted into a simulated shelf configuration, and competing packs can even be included for comparison.

Sometimes, the client's marketing staff want to be closely involved with the design process at all stages. A convenient way of achieving this is for the marketing contact to sit alongside a designer in front of a Paintbox-type computer system. Within the space of about a day, the designer and client can then work closely together by exploring various concepts on screen, changing colours and design elements as necessary. Logos can be increased or reduced in size, and an almost endless variety of experiments can be conducted, under the client's instructions, but guided by a senior designer. This can be a cost-effective method of obtaining the desired results. The output can be in the form of 35 mm slides or prints, which can be taken away for subsequent discussion. Technology also allows this process to be carried out without the client and designer being at the same physical location.

Any potential design solutions must be ranked in order of cost. As soon as the design directions have been established, a member of the design team must begin to consider the cost factors, to the stage of being able to indicate that design A will be a more expensive option than design B or design C. This aspect of the design process is important when technical assistance is at hand to advise on costs and practicalities. An in-house technical department is a distinct advantage for a design consultancy. This function is normally carried out by the 3D designers, who, because of the more technical nature of their art, are in a position to know what is feasible and what is not, or to have access to the relevant sources of information. This applies also to printing, when the limitations of each type of print process needs to be known. A consultancy which offers in-house artwork production always employs a print technician with the expertise to advise on these aspects.

In a design studio, the first impression may be of a chaotic environment, with books, magazines, pack samples, sketches and 'bits' of current and previous design work scattered around. There may be a combination of computer terminals, drawing boards, tables, cutting areas, spray booths and materials stores. Attempts to create a tidy working area seemed to be doomed to failure within a design environment, with tidiness being at best spasmodic. Creative activities seem to necessitate a degree of untidiness, and time constraints do not encourage excessive clearing up. However, a closer inspection will reveal the background information that the designers need. Designers cannot operate in a vacuum. They thrive on ideas obtained from a whole spectrum of material. This essential stimulation may come from reference books, some of which are concerned with design, and others with the subject of the project. Cookery books, manuals of dog breeds, and books on horticulture and plants, architecture and just about every other topic form a reference library from which the designers can extract specialist information. The library will include design books that look back over classic designs (both packaging and nonpackaging), photography books, and books about art, anatomy, interior design, lamp design, cars, furniture and many other subjects. Women's, men's, fashion, design, computer-graphics, marketing, technical and artistic magazines will provide the designers with up-to-date general information and information on the subject of the project. Designers need to experience the mental stimulation provided by such material as much as they need to keep abreast of contemporary trends in fashion and lifestyles. Graphic design also calls on artistic skills, for example when it is necessary to include illustrations of fruit on a pack. In this case, there will be a need to refer to actual fruits or photographs of fruits to ensure accuracy.

While concepts are being developed, many designers will want to work directly in 3D format. This will certainly be the case if the project includes structural design, but it is also often the case if the project is confined to graphics. Developing a label for a beer can, for example, is a graphics project which most designers would approach by initially working in the flat, and then wrapping the design around a can. The consequent distortion often leads to unplanned visual effects, to the detriment of the design concept. If the effects of distortion are taken into account from the start, perhaps by conducting some experiments working directly onto the can, new possibilities can open up, and errors caused by working only in the flat can be avoided. Much 3D work, even at the conceptual stage, is carried out using rough models. These are strictly for the benefit of the designer, and are created for the purpose of allowing ideas to be worked out. Foamed plastics are often used, as blocks of this material can easily be shaped by hand or machine sanding. It is often quicker to work in this medium than to produce perspective drawings, and there is the added benefit that models assist the designer in working out what happens to grooves and other features when they are moved from one panel to another. If designers use only drawings, they can cheat and ignore how such features and compound curves, for example, will work out in practice. Modelling clay and card are frequently used to help designers think through ideas, with many creations ending up on the workshop floor before a design route with true potential is identified.

Several computer systems allow 3D modelling on screen. The model can be rotated and viewed from any angle, coloured, textured, and lit from any point. Glass and chrome metallic surfaces can be realistically reproduced. Most systems allow a wireframe model to be created. The surface of this is then rendered to the designer's requirements. This sounds like an ideal method of producing conceptual 3D models, but it has its drawbacks. Unless the object has a very simple and regular shape, drawing it on the screen can be a time-consuming task. Cylindrical containers are quick to reproduce, but the incorporation of an asymmetrical feature is a task that is by no means unsurmountable, but that requires familiarity with the software, which takes time. For some complex designs, it is much quicker to make foam models, as several of these can be completed in the time that it takes to produce one computer-generated model. Rendering a wireframe model can also be a long process, taking hours of computer time in some instances. This is changing as new systems become available, and software becomes more intuitive and faster. Where computer systems are currently useful is after the rough-concept stage, when the design routes have become established, and highly finished models are required. The standard of presentation of a computer model can be as good as a photograph of a finished pack.

The first stage of a project is by far the largest stage. In minor projects, it may be the only stage, but in more complex projects, it may the first of many. It is during the first stage that the creative thinking is done, and design routes are established. During this process, constraints are gradually introduced when thought appropriate by the design director. These constraints can be due, for example, to costs or the manufacturing process. The team should then be in a position to make their recommendations as to which designs have the greatest potential. It is essential that an in-house session that involves all the participating members in the team be held to review the work, and to agree recommendations. This entails setting a cutoff point and calling a halt to further work before preparing for a client presentation. The team must make recommendations, and not merely show work to clients, judge their reactions to the designs, and recommend those that the clients appear to like. When the meeting has been held, and the design routes to be recommended have been agreed, the studio must prepare for the first-stage presentation.

PRESENTATIONS

The purpose of a presentation is to communicate to the client what work has been conducted, which designs are favoured, and why these have been selected while others have been rejected. The client is entitled to see all the work done, and to be told the rationale behind the designers' thinking. This is primarily a communications exercise. A poorly presented case for a pack form will have little impact, and so it is worthwhile structuring the presentation format so as to ensure that it is positive in tone without being dogmatic.

The preparation of material for presentation is important. The choice of format is influenced by the nature of the project, the costs, and the timescale for preparing material. Whether the presentation is to include models, visuals, on-screen computer-aided design work, photographs or other formats, the standards of each must be compatible, so that fair comparisons can be made between the designs.

Presentations must be structured, leading the client through the development of the study in a logical and easily understood manner. Starting the presentation with a recap of the project brief sets the scene. Showing examples of the store survey, using selected slides to illustrate particular points, helps to establish the parameters relating to the point-of-sale conditions and the competing products. These are the areas that are deemed important by the design team, and they must be highlighted, and the rationale explained.

After the store-survey review, other significant information from the predesign research can be introduced, such as the outcomes of discussions with other agencies, and the results of plant visits. The client will then be eager to get to the main objective of the study, and see the design work. If any interim meetings have been held, the client company will have a fair idea of what directions have been taken, but they will not have seen the latest work. Not all of the client party will necessarily have been involved at the interim stages. Therefore, a complete review of all work done will be required, even if interim meetings have been held. In this case, the review should at least recap the work done before the continuation of work after the interim meeting.

There is no substitute for showing 3D packs during a presentation, even in a graphics project in which new graphics have been applied to existing packs. Consultancies use various techniques at different times to show work. Standard methods are to reveal the new designs by using the 'rabbit out of a hat' technique, or by removing screens over packs placed on a shelf. Whatever method is used, the presentation should follow a logical progression, showing designs, explaining why the design routes have been selected, and explaining why they should be promoted or rejected. Recommendations should be left until the end. Occasionally, there may be a 'joker in the pack', i.e. a design that is extravagant or that breaks the brief apart, but that has a particular merit. This can be introduced to stimulate debate, but not put forward as a serious contender, unless the designers feel that it is one. There are times during the course of a study when the design team may challenge the brief, if they feel that an area has been overlooked, or that the writer of the brief is mistaken in some way. There is no harm in designers moving outside the brief and exploring possibilities, provided that they return within its boundaries in due course, or provide a reason for further questioning of the brief. Time spent in areas outside the scope of the brief must be kept to a minimum. Clients will pay for what has been agreed, and will not take kindly to paying for work done in other areas.

Communications must be mutually understood if they are to be successful. A presentation should not be a diatribe by the designers, but a two-way meeting in which the client is involved, and can participate in the discussion. The client may simply sit through the whole meeting, and agree to proceed with the recommended pack. However, there will usually be some criticism, praise or comment which must be addressed by further work. After a presentation, there must be a debriefing meeting with members of the design team to decide future actions and allocate tasks. The second stage of the project, if there is to be a further stage, will probably have to be defined in a working document, worksheet, or other form of studio instruction. In most projects, the timing can be so tight that the designers do not have time to wait for such a document to be issued, and they will start work immediately on the basis of verbal instructions. Design stages sometimes become blurred when time constraints pressurize work. Paperwork can be done later; it is the design solutions that are important. When time constraints are tight, which is the case in about 90% of projects, all the parties are keen to press on, and they accept the fact that the paperwork will lag behind. At the end of the project, however, when the paperwork is caught up on, particularly the invoices, there are often all sorts of disputes, for example about costs, and whether people have or have not been advised of work. Thus, if proper documentation is not issued throughout the project, it can become unpleasant at the project end, for all the parties involved. Therefore, despite pressure of work, each stage should be fully documented and approved before, or shortly after, its start. In particular, clients must be advised of the costs of proceeding with the project, especially if timescales are being telescoped.

7
The 3D-design process

The essence of strategic packaging design is a planned, strategic approach. This applies in the same way to 3D design as to other aspects of design studies. The study must proceed through a series of stages, each of which builds on the agreed directions identified in the one before. There may be any number of stages, depending on the complexity of the project, but a project typically has three:

- *Stage 1:*
  - brief,
  - store survey: product position at point of sale, and competing products,
  - production site visit: discussions with production department, engineering, machinery, storage, transport: opportunities and problems,
  - advertising agency: discussions and viewpoints,
  - current-pack analysis: strengths, weaknesses and costs,
  - design concepts: exploration of a range of materials and pack forms,
  - interim meeting: rejection of unsuitable ideas,
  - further work: sketches, CAD, rough models, cost rankings,
  - presentation of recommendations.
- *Stage 2:*
  - further design work on agreed design routes,
  - model making,

  - outline specifications,
  - discussions with purchasing department,
  - discussions with suppliers: outline costings and modifications,
  - research models (if required).

- *Stage 3:*

  - final modifications,
  - final drawings and specifications,
  - ongoing supplier liaison: prototypes,
  - production trials, filling trials, and transit tests.

The above schedule is based on the development of a pack style for one size variant.

The first stage is usually the longest and most expensive stage, during which much time is spent in gathering information before design concepts are considered. Usually, a brief is presented to the design group by the client's marketing staff, who are responsible for overall control of the project. Typically, this is talked through, relevant samples are shown, and a written summary of the brief and backup information, market research etc. are left with the design studio. Occasionally, the marketing personnel will not have originated the brief at this point, and it may be the task of the studio to extract the details and commit them to paper for subsequent client approval.

It is unlikely that designers will be familiar with every type of product presented to them by prospective clients. Even if a product is in a category of which the designers have prior knowledge, the market changes quickly, and an update will be required. The most effective method of conducting this is to carry out a store survey, and visit a selected range of outlets where the product is likely to be stocked. If the product is to be sold in other countries, a member of the design team must visit these countries, and take photographs of product displays for the use of the design team. Photography is a useful tool for recording product displays as they actually are, and not as the client's marketing staff think that they should be.

Few people would fail to recognize a Perrier, Coca-Cola or Michelob bottle at 20 m. One does not have to read the label, as one can immediately identify the brand by shape alone. This is what 3D design can achieve: the development of a unique brand identity without recourse to graphics.

All packaging is 3D. The packaging on supermarket shelves has a vast range of shapes and configurations: cans, cartons, tubes and bottles. Some of these identify the type of product by their shape, and others, such as the Perrier bottle, actually identify the brand. The packs on the shelf also demonstrate another, more subtle, aspect of 3D design: the incorporation of features to provide consumer benefits. Pouring spouts, easy-open features, peelable lids and resealable tops are all examples of 3D design at a secondary, more technical level. The selection of the materials in packs is also interesting. Combinations of materials can maintain the shelf life of the product, but also allow the product to be cooked within the pack.

SHAPE

Armed with a pencil and a blank pad of paper, one might create large numbers of shape suggestions. One would soon want to know what the shape was to be used for. If, for example, the shape was for a jam container, one would immediately have parameters to work within. One's experience, as a consumer of jam, would tell one that it was necessary to ensure that the product could be accessed. One might consider how the jam was going to be inserted in the pack, and how the pack was going to stand on the shelves. One's knowledge of existing pack forms (in this case jars) would also colour one's design judgment. To move the design on, one would need further information. The problem would have to be defined before one could attempt to produce a design solution.

Designers must know about:

- the nature of the product,
- product imagery,
- how and where the product is to be sold.

Much more information will be needed in due course, but these basic requirements are considered first.

NATURE OF PRODUCT

If a designer does not understand the product, he or she may waste time in indulging in flights of fancy that have little relevance to the product. The client's marketing staff have a responsibility to provide product details, and designers have a responsibility to develop an understanding of them. It is often worthwhile for a member of the design team to visit the production site, see the product being manufactured, and discuss the product with those responsible for its formulation. Designers need to know how the product will be filled, and what protection will be required to maintain product integrity throughout the required shelf life, before and after purchase. Information will be required on the product's tolerance to moisture, microbiological contamination, gases and odours, temperature, light, and the mechanical hazards of dropping, crushing, vibration etc.

PRODUCT IMAGERY

Imagery is an abstract concept which must be translated into physical 3D form. It is probably the most significant driving force that propels design work in one direction rather than another. However, because of its abstract nature, it is one of the most difficult qualities to communicate about and exercise judgment about. In the example of jam, the marketing strategy may require a pack which suggests that the product is home-made, that is redolent of 'old England'. This imagery is focused on bygone values, rather than contemporary ones. Under these criteria, a PET jar may be inappropriate because its sophisticated construction has little in common with the required imagery. The designer may research the question of what sort of shapes reflect the imagery values specified by looking back at previous generations of products which were considered to have the right qualities. The designer may also look at current packs that have the same type of imagery, and so, in the jam example, 'traditional' ales, mustards and pickles may provide clues as to what shapes consumers associate with 'traditional' values. Interestingly, the age profile of the potential purchasers may show that they will be too young to remember the original packs on which modern 'traditional' packs are based. Not many jam purchasers have ever bought 'real' jam with a gingham-cloth lid, and yet this design has been widely adopted as a graphics device on jam lids, and it is somehow associated with jam which 'grandmother used to make'.

One of the best ways in which the client's marketing staff can communicate their required imagery concepts is for them to produce a profile of the purchaser. In the example, discussed in the first chapter, of a very expensive new brand of premium motor oil intended for high-performance cars, the marketing department knew that owners of expensive high-performance cars did not buy oil, as the cars were almost always serviced by dealers. The market for the oil comprised owners of 'hot' hatchbacks that were three or more years old who loved their cars, would not trust garages to service them, and could not afford to pay garage rates for tasks which they were capable of carrying out for themselves. It was difficult for the marketing department to communicate the imagery that they wished to see built into the pack, and a full explanation to the design team was required to get the message across. The marketing staff were able to paint a vivid verbal portrait of the typical consumer of the oil as being male, 25–50 years old, a three-year-old Ford Escort XR3 owner, and the possessor of a detailed knowledge of car makes and specifications, but limited mechanical competence. The typical consumer did not have a garage, and worked on the car in the drive or out on the street. He watched motor sport on television, but did not participate in it, and rarely visited motor-sports venues. If he had had the money, he would have bought a new BMW (white). He took package holidays abroad, and drank lager.

Such a description may contain more detail than is required. In the example, the target consumer might have been fictional, or based on fact, but the designers formed a very clear picture of him and the sort of imagery which motivated him. They even called him Kevin. A key phrase was 'he loved his car', and another key piece of imagery lay in his aspirations to own a BMW. The pack had to reflect 'BMW' rather than 'Ford Escort'. This also had implications for the graphics. The resulting pack shape was essentially practical; handling was easy, and the pack offered nonglug pouring. It also incorporated up-market car imagery; it had an engineered appearance which appealed to the target market fairly successfully.

While shape is an ideal medium for conveying imagery, the selection of the ideal-image lead pack may need to be modified by more practical constraints.

POINT OF SALE

As most products are sold in self-service outlets, the physical location of the product requires careful consideration. Few manufacturers are in a position to dictate the location of the product to any of the major multiples. It is the merchandising section of the multiples that decides where products are located in the store. Designers must see the point-of-sale area for themselves. Many consultancies visit a selection of the outlets where the product is to be sold, and photograph the appropriate point-of-sale area, taking particular note of the lighting levels, the competing products, and the products that will be displayed above, below and to the sides of the product. They may also count facings, i.e. the numbers of packs on the shelf categorized by brands, product variants and product types.

It is no longer enough for designers to produce interesting shapes that incorporate the desired imagery. Packs must conform to the multiples' requirements on profitability in terms of return per length of shelf. The pack configuration must satisfy the merchandising needs of the stores. Either the client's marketing department must supply this information, or the design team must contact the retailers to obtain it. The pack may then have to be modified to suit the shelf requirements. This has the effect of standardizing shapes and making the job of the designer, to create a distinctive shape within fairly strict dimensional restrictions, more challenging. If these restrictions were strictly adhered to, all packs would be brick-shaped, and the possibility of distinguishing brands by shape alone would disappear. Fortunately for design businesses and the interests of the consumer, there is still scope for shape to play a major role in design.

TARGETING MARKET SECTORS

In new-product developments, it may be possible to create a distinctive shape which creates its own market niche. The first liquid detergent that incorporated a dosing ball on top of the pack had a distinctive shape on the shelf that was quite unlike that of any other product. Together with a major advertising campaign, this pack soon created an expanding market for liquid detergents, with the dosing ball being featured as a device that ensured that the product went 'right to the heart of the wash'. Apart from being the basis of a clever marketing strategy, the pack form itself was unique, creating an enviably unique brand identity. As with most successful brands, it was not long before competitors started to copy both the product and the pack style. Other brands and own-label products now sit alongside the original pack. The product and the pack created a new market sector. They initially dominated it, but were then forced to share it as other brands caught up.

Within the hair-care market, there are market segments in which the type of the product is communicated by shape alone. Antidandruff shampoos currently tend to follow the Head & Shoulders pack design by incorporating an elongated neck section. Those manufacturers that choose to enter this market are almost forced to follow the market leader in order to compete. Consumers purchasing antidandruff shampoo often look for white bottles with long necks on the shelves, and then make their selection on the basis of brand preference and price. It is increasingly difficult to adopt any other pack configuration and still be competitive. If a product is first in the market with a distinctive pack shape, it can establish a market sector. This does not guarantee brand supremacy for ever. If the product is successful, others will surely follow, but there is a time slot of dominance while the competitors struggle to catch up.

Toilet Duck is another example of a new pack shape which moved into an established market and soon dominated it because of its distinctive shape. The swan-neck configuration provided an immediate point of difference between it and more conventionally packed competitors, and this created a focus for advertising. The shape clearly indicated how the product could be used effectively. This revolutionized the market, and achieved brand leadership for its originator, although this was then gradually eroded by 'me-too' products and packs.

SIZE IMPRESSION

The ability to manipulate shape provides the opportunity to create size impression or maximize the area for graphics. The opportunity to maximize the size impression must be treated with caution. Packs must not be misleading. Apart from the fact that this may attract the attention of trading-standards officers, it is bad practice, and ultimately bad business, to produce packs which are deceptive. Consumers may be fooled or disappointed once, but repeat purchases will decrease, and, even worse, the brand may be tarnished in the mind of the consumer for a long time. While the size impression must not be overlooked, it should not be accorded overriding importance to the detriment of all other design issues. There used to be a trend in toiletries packaging of tall flat packs, providing a large area for graphics. The packs were so thin that stability became a problem, both in filling and in the home. Unstable packs are very annoying for the consumer, and are a reason for them to switch brands. This is also the case for packs that incorporate overlarge closures. Often, stability is affected in this situation, particularly when the pack is less than full. The large cap may also be perceived by the consumer as misleading, causing a negative reaction. The designer must therefore use constraint, and balance practical issues against the size impression. There will always be pressure to increase the available area for graphics, but this should never be done at the expense of competent 3D design.

There are some situations in which it may be valid to place more emphasis on size impression. Aerosol packs contain propellants, and, if the pack is changed to a nonaerosol pump format, the new pack will be smaller. To the consumer, the diminished pack size will apparently offer less product for much the same money. The pack will also have to compete with other brands of similar products which remain in the aerosol format. One solution to this problem that has been used is to waist the container, leaving a full-width shoulder at the top and tapering sides. The graphics area is reduced, but the shape becomes more distinctive, and the product is moved away from the pure cylindrical shape that is associated with aerosols. Rather than attempting to compete head-on with aerosols, such new shapes provide an opportunity to reposition the product, possibly by emphasizing the virtue that nonaerosols do not contain chemical propellants.

The same problem occurs when materials are changed from glass to plastics. Thin-walled plastics containers appear to hold much less than their thicker-walled glass counterparts. For a similar size impression to be maintained, the container configuration may have to be rethought. When the original glass shape is well established, and intimately associated with the brand, the task becomes difficult, and the problem may be sufficiently great to rule out any change of materials. The original, glass-waisted Coca-Cola bottle shrinks dramatically if the product volume is maintained in a plastics reproduction. This classic design therefore does not lend itself to imitation in plastics. The switch in materials has to be more significant, or other strategies have to be considered. The use of PET bottles for the 1.5 l size, and cans for smaller sizes, avoids the problem, but graphics rather than shape are relied on, to communicate the brand.

Packaging-materials changes can, as well as posing size-impression problems, also provide opportunities to establish brand identity through the medium of shape. Mazola used to use a standard cylindrical glass bottle for its corn oil. A change to a plastics bottle would have resulted in the apparent volume reduction mentioned above. Corn-oil imagery, in the form of raised corn kernels, was incorporated in a new plastics bottle, which, as a result, had much the same external dimensions as the previous bottle. It was, however, far more distinctive. The graphics area was reduced to a smaller band at the centre of the container, which, because of the brand differentiation achieved through the 3D design, was acceptable.

Rather than aiming for pure size impression, it is better to consider designs which are distinctive, remembering always that pack standout is one of the most valuable contributions that 3D design can make. Manipulation of shape is probably the most effective method of achieving a unique brand identity.

CONSUMER BENEFITS

3D design provides an unparalled opportunity to build consumer benefits into packaging. Some of the most important benefits relate to:

- opening, closing and resealing,
- the dispensing of contents,
- ease of handling,
- opportunities to see and feel the product before purchase,
- the in-pack cooking of food,
- after use and ease of disposal,
- tamper-evident packaging, security, and product-integrity features.

All packaging design should consider the end user, and how to make the tasks of opening, dispensing and using the product as easy as possible. Everyone has experienced the frustration of trying to open packs, often spilling the product as a result. Most consumers have their own pet hates, but, in a recent newspaper survey, gable-top and brick-shaped milk cartons received the most criticism. (Biscuit wraps and audiocassettes came close seconds.) Surprisingly few of the respondents seemed to have kitchens that were equipped with scissors to ease the task, although all of them had can openers and corkscrews. It seems that consumers are quite prepared to use tools to gain access to wine and canned goods, but not to products in cartons, flexible packets or wraps. However odd the logic may be, most households struggle on with opening these types of pack without any mechanical assistance, and this has been true for a considerable period of time. Despite this information, packs are still produced which are a challenge for the end user to open. Opportunities exist for improvement, and for some brand preference to be expressed as a result of easy-to-open packaging. The above questionnaire also surveyed biscuit packs with built-in tear strips. Many of the respondents had not noticed them, although, when the respondent did use the tear strips, these were highly praised. This is a consumer benefit that is largely lost, simply because it is not evident, resulting in a lost marketing opportunity. Consumer benefits must be obvious, preferably before purchase, when they can help to move products off the shelves. Subtle, hidden devices may require an elaborate explanation (which is unlikely to be read), and they may improve the performance of the pack, which is good, but not be perceived as a benefit by the consumer.

Many products, particularly foods, come under the category of convenience products. In one way or another, they are formulated and/or packed to give ease of use after little or no preparation. With busy lifestyles, this has become a major factor in marketing many types of food, and so this market sector is important. In some products, the convenience product and the pack are inseparable, and together constitute the total product. Microwaveable food is one example. One product that is marketed in the USA, the home of convenience foods, offers a complete breakfast in one pack. By the clever use of susceptors (boards selectively coated with metal at strategic positions), a frozen meal can be placed in the microwave oven for just a few minutes. The orange juice is thawed, the coffee is heated, the bacon is crisped and the 'hash brownie' potatoes are cooked. Whether the results are culinarily acceptable is largely a matter of taste. However, the pack and product coexist to provide the maximum consumer benefit.

Most consumer benefits that are built into packaging are more modest, and consist of ways of making the pack more user-friendly. The designer should try to assist the end user, and enhance the pack to gain a marketing advantage over competing products. The move from bags of flour to rigid packs, for example, allowed a sprinkler portion to be incorporated in a plastics lid. This was not essential, but was an added benefit for the consumer. The same principle applies to nonfood products, even those that are not sold to the general public. For example, several formats are used for packaging dental drills. The most favoured designs present the drills to the dental technician in an organized fashion, using a carousel type of dispenser. While the technician may not be responsible for purchasing drills, his or her views, which are based on experience with various pack forms, may influence the purchasing decision. The ability of nursing staff to influence brand selection within the field of medical packaging is recognized by pharmaceutical and medical-equipment companies, who actively pursue methods of making their packaging easier to use than that of competing products, despite the fact that the responsibility for making purchasing decisions may lie elsewhere.

It would be impossible to list all the consumer benefits that can be obtained from differing pack forms. Serious thought must, however, be given by marketing staff and designers to all packaging design, to assess whether there are opportunities to help the end user of the package, and thus add value to the packaging.

MATERIALS SELECTION

It is usually fairly pointless to attempt to design packaging without first having a good idea what materials will be involved, and having a basic understanding of their properties and production limitations. The basic categories of material are as follows:

- wood,
- paper and board (including corrugated board),
- plastics,
- glass,
- metals,
- textiles.

Packs may also be composite structures that combine two or more of the above categories of material, or occasionally less common materials such as straw, cork and ceramics.

It can be a mistake to dismiss out of hand materials that are apparently unsuitable. All materials should be given serious consideration, and then promoted or rejected. An aerosol of dairy cream may not be an astonishing product, but when it was first launched into the UK market, it was revolutionary. No-one would have thought of packaging cream into an aerosol format before, as it was such an unlikely choice of pack form. The success of the product owed a lot to the boldness of the designer and the imagination of the marketing team.

In most cases, it is the nature of the product which dictates the choice of materials. The need to maintain product integrity against degradation due, for example, to the ingress of moisture and gases is a primary factor. Other factors may be a requirement to open, close or reseal, or the ability to use existing machinery for filling. The essential conditions must be fully documented, so that the design team can consider materials in the full knowledge of what is required.

The choice of materials must be made first on structural grounds. The pack will have to contain and protect the product through all the stages of production, storage, transportation, sales distribution and end use. Second, imagery must be considered, and the materials must be matched with the required product image. For a traditional product, such as a full-fruit jam, glass is a strong contender. The design of traditional products looks backwards to older pack forms. Exclusive products, for example a £2000 watch, may be best represented by wood, leather or metal packaging. Most FMCG products are neither traditional nor exclusive, and so there is considerable scope in the selection of materials. For a shampoo, for example, it would be unthinkable not to consider plastics. Containers are most likely to be blow-moulded or injection blow-moulded in polyethylene, polyvinyl chloride (PVC), or PET. Unless the designer is aware of the limitations and production parameters of these processes, time can be wasted in producing designs which are not easy, or even impossible, to manufacture.

As one example of the principles of making choices about materials, consider the packaging for a premium motor oil. Suppose that the brief states that the oil is to be packaged in 5 l containers, and the imagery is slanted towards a performance product for high-performance cars.

Wood is obviously inappropriate as a packaging material. While glass has been used to package motor oils, it has only been used for smaller-sized packs. A 5 l glass container is not practical. When board is considered, there are some possibilities. Foil-lined 1 l cartons of motor oil are available now, and so the problems of extending the size to 5 l should be considered. Would 5 l of oil be too heavy for a foil-lined board pack? How might it be filled? Would that type of pack be appropriate for a premium motor oil?

The answers to these questions lie partly within the brief, and partly in the judgment of the designer. When an answer is in the negative, it need not be pursued, but, when there is some degree of doubt, it may be worth a second look to identify the advantages and disadvantages. It is important that all materials be considered, and that there is a rationale behind rejection decisions. Designers must not be too quick to reject the unusual. There could be an element in these ideas which could be adapted to give exactly the pack required. In the example, if a foil-lined carton is not up to the job, a variant of it, the bag-in-a-box concept, would be considered. This requires a combination of materials: board and plastics. Some liquids are already packed in this way, and so there is a good chance that oil can also be packed in a similar fashion. The advantages and disadvantages of this idea must be examined.

The designers must consider the environmental advantages that the bag-in-a-box pack offers. It uses less plastic than a rigid container, and the board outer is recycleable. This will need further investigation, but there is a potential environmental benefit which could provide a marketing advantage over competing packs. The compact size of this pack is likely to provide benefits in storage and distribution, and the cuboid shape will make point-of-sale display easier. No corrugated fibreboard shipping cases will be required, as would be the case for, say, four 5 l conventional plastics bottles of oil. The decoration on the box can be of a high standard, with a litho-laminated liner being used, which can be printed over the whole of the surface area. In use, the bag will collapse as the oil is dispensed, preventing the entry of air through the pouring spout, and consequently allowing 'glug-free' pouring. To provide consumer benefits, the spout could incorporate a tap to regulate the oil flow, and the outer box could be equipped with a slot to show the level of oil within the pack.

While this concept is technically feasible, there are some possible negative aspects to it. It might not be possible to fill the pack on current oil filling lines, and new equipment might therefore be necessary. With such a premium product, the filling rates are generally not high, and so this may allow some scope. The imagery of a cuboid container might not match the association with fast cars, as a cube is not the most exciting possible shape. The overall appearance could be improved by a careful selection of materials, and perhaps the use of foils or metallic inks. The graphics would have to provide the correct degree of imagery, and the pack form would maximize the area for graphics use. The cube shape would be compact, and therefore decrease the size impression. However, it would probably be sufficiently different in terms of configuration from competing packs for this to be of minor importance; it would be necessary for the marketing group to consider this issue. Oil packs need two types of handling aid: one to enable the pack to be carried from the point of purchase to the car, and the second to assist in pouring oil into the engine. A strap handle, perhaps in conjunction with an inbuilt cutout on the box, could be designed to fulfil both these roles.

It is evident that this is a materials combination that has some suitability for the task in question, and that it has enough potential to justify a more detailed investigation. Other materials, for example, plastics containers and metal cans, would need to be examined in the same way. In each case, a strategic packaging-design approach should be used. This entails systematic examination of the options, and the evaluation of their potential.

To examine a broad spectrum of materials, the design team needs to be familiar with the nature of the materials, how they can be used, and what processes are involved in their use. It is unrealistic for designers to have an in-depth knowledge of every packaging material. What they need is an appreciation of what is and is not feasible, and a knowledge of where to go for more detailed specialist information.

ROLE OF SUPPLIERS

Many suppliers have their own design facilities that specialize in their own particular packaging media. Container suppliers, in particular, often have sophisticated CAD systems which cannot only assist in the design process, but may also be linked to toolmaking equipment, i.e. CAD/CAM systems. While it is possible to use suppliers as a primary design source, design is not their forte. They do not have the resources to research product areas in the same way that specialist design studios do, and they cannot use the strategic approach to packaging design that is advocated in this book. Nevertheless, as a secondary design source, their detailed knowledge of materials and processes is invaluable. Ideally, suppliers should be involved at the design-concept stage, so that they can evaluate design candidates from a manufacturing point of view, and thus make a positive contribution to the study. Suppliers can advise on the selection of materials, and the best way to interpret and/or modify the design to achieve cost-effective results. They have the final responsibility of producing the pack, and so their contribution is important, and their views should not be ignored. Often, however, production constraints are given priority, to the detriment of the preferred design. In such cases, suppliers must be brought into the project at an early stage, so that the thinking behind the design routes can be fully explained to them. Suppliers are receptive to new ideas, and, if they are part of the team, they become enthusiastic about pushing their technology one step ahead. It is a mistake to enlist their help only when the project is already too far down the road for changes to take place. An early involvement pays dividends.

Suppliers can provide the following:

- materials and processes expertise,
- design modifications to facilitate production,
- details of costs,
- production-scheduling information.

It is usual for supplier involvement to occur when the initial design concepts have been established. Sketches, drawings and probably rough models will have been prepared which the supplier can then see and evaluate. The supplier's role will be to assess the practical issues that the designs raise, identify problems, suggest solutions, cost out the pack in terms of tooling costs, and parts prices, and establish the timing parameters.

When the pack production is dependent on tooling, as is the case, for example, with a blow-moulded bottle, the projected annual pack quantities will have a direct bearing on the type of machinery required for production, and hence the tooling required. Tooling is expensive, and it may add months to the initial leadtime. Close cooperation between the design team and the suppliers is essential to cope with problems as they arise. In a complex project, this is invariably the case, and detailed modifications may have to be made to the design for production reasons. As the guardian of the design, the pack designer must ensure that the design features agreed with the client are not compromised, and yet be prepared to listen to the supplier's needs and adapt the design when this is essential.

It is an illuminating experience for everyone when packs are displayed in real conditions, and not simply placed on the studio shelves or the marketing manager's desk. The main panel, which has been carefully crafted by the designer, may not be visible on the store shelves at all. Flour bags are one example of this. It is often the base of the pack that is displayed when store staff lay packs on their backs to aid stacking. The retailer will probably not change its display techniques, and so the onus is on the design team to realize how the packs will actually be displayed, and react accordingly.

A store survey will identify what packs are likely to be sold above, below and next to the designed pack. All these will compete for the shopper's attention, and, while it is important to know what the competition is doing, it is therefore also important to establish what other categories of product will be in proximity to the pack. From the 3D designer's point of view, the new pack must maximize the product's standout from all the surrounding packs, through the use of shape alone, or by the use of different materials or pack forms.

Further details about predesign research have been given earlier. The importance of the store survey (i.e. seeing what is actually happening at the point of sale) must be emphasized, as this provides the basis for accurate design work.

Because standout is the first priority for all packaging, 3D design plays a vital role in effective design. The incorporation of consumer benefits can make the pack perform better than its counterparts, and provide a reason to purchase in the face of the competition. The selection of materials for construction is part of the design process, and it is important in terms of ensuring the pack's success. The pack must not overpromise in terms of size impression. It must perform its role satisfactorily in every respect. Using the required product imagery, the audience can be carefully targeted, and the pack shape and texture must therefore portray exactly the image that the client wishes it to. With 3D design, there is, above all, an opportunity to establish a unique brand identity for the product, to ensure that it becomes a brand leader, and to provide a tool for advertising purposes, that most powerful persuader to purchase.

8 Graphic design

More has been written about graphic design than about any other topic in packaging design. For many people, packaging design is graphic design. It is also a controversial area, particularly within the design community, in which there is much discussion about what constitutes good graphic design and what does not. This is often a matter of taste, with each individual having a different concept of what good design taste is. While it is easy to be critical of the graphics on a pack, the criticisms are, more often than not, made without any knowledge of the brief. Any designer, or anyone else for that matter, can pick up a pack and analyse its design content without knowing what the design's objectives were.

The aims of strategic packaging design are simple, although how they are achieved depends on the talent of the design team. All that is required is a design that sells a product and promotes a brand. If the result is also a classic design, or one that is admired by other designers, this is a bonus. It is not necessary to produce a work of art, or designs that massage their designer's ego. Packaging design is a commercial tool, the application of which must yield successful results as measured by sales. This is one of the reasons why some design consultancies do not enter design-award schemes. The managers are anxious that their designers should not be distracted, and not focus on design issues rather than results for clients. This may seem a harsh policy in which artistic merit is demoted in favour of commercial success. However, it reflects the reality of the packaging-design business. This does not mean that packaging design has to be harsh, ugly and without design merit. Many successful designs are cleverly crafted, possess considerable qualities of artistic excellence and elegance, and display their designer's talent.

Most graphic designers working within the sphere of packaging design have an art-school background and a training in graphic design. Only a few have training specifically in packaging design. When a designer has learned the techniques of design production, and has a concern for the more aesthetic aspects of design, it can come as a shock to him or her to be faced with the commercial realities of the packaging-design environment. What young designers can contribute, however, is a freshness, an uninhibited approach, and an empathy with contemporary values, lifestyles and fashions. More experienced and senior designers, who have elected to specialize in packaging, will have achieved an understanding of the packaging issues, the markets and what sells products.

Standout is the highest-priority design quality. If the pack is not noticed in the time in which it takes a consumer to scan the shelves, then it will remain unnoticed, and unsold. Standout is the vital ingredient that all graphic designs must have, in terms of attracting attention, brand recognition, product-type identification, and differentiation from competing products.

It can be an illuminating exercise to select and examine a product area within a supermarket environment, assess which packs are first seen and which brands are immediately recognized, and then try to determine why this is, and what the design has that makes it stand out. One might wonder why, if standout is so important, fluorescent inks or garish colours are not used more often. In some market sectors, they are, notably in that of detergents. Designing for sales is always a compromise, and, while standout is first on the list of priorities, it is not the only consideration. Branding and imagery are also vital. Building a brand represents a considerable investment, and a brand leader is probably one of its manufacturing company's most important assets. Consider how many millions of pounds must have been spent on promoting the major brands of detergents, for example Tide, Ariel and Persil, over the years during which these products have been competing for brand leadership. The investment is colossal. Any design which diminishes the brand identification of any of these products is a nonstarter, no matter how 'tasteful' the design may be.

The correct imagery is vital to product success, and graphic designers play a key role in this area. Designers are not only paid for their graphics skills, but also their ability to reason. Creativity is important, but the designer must also be able to direct his or her creativity towards achieving the desired result. The ability to pick up imagery clues, interpret them, and produce designs which emphasize them is a skill that is usually developed and finely tuned over time.

It is thus more effective to use a design team than one individual. The freshness of the junior members of the team coupled with the experience and understanding of more senior designers helps to produce balanced solutions to design problems. The ability to work in a team is something that cannot easily be taught at college, but is essential in a design studio. Effective management of the designers, who can be a difficult group of people to organize, is also essential. Each designer will probably hold strong views on his or her own interpretation of the brief, and defend his or her corner enthusiastically. At times, the design director will have to curtail some design experiments and promote other directions, at the same time as attempting to keep the team focused on a common goal.

DESIGN INFLUENCES

Graphic design seems to be more quickly influenced by current trends than are other design disciplines. This is perhaps because it is a transient form of design. It is relatively inexpensive and easy to change graphics at frequent intervals. Product designs, furniture designs, automotive designs and architectural designs are longer-lasting; they have to be, as they represent a much greater investment in terms of tooling, time and money.

Design originality is hard to substantiate, as nearly all designs are based on concepts that have been used before in one form or another. Designs may draw on concepts from art and design produced in other countries. Most studios will have examples of Japanese, American and European graphics work. It is interesting to see how other countries, perhaps with very different lifestyles, approach design. Values may differ, and cultural differences produce interesting product variations. For example, the way in which a Japanese designer approaches the design of a soap pack may not be strictly applicable to the task in hand, but it is a new design dimension, and it provides further food for thought. Inspiration rarely stems directly from the analytical part of the brain. It occurs when the thinking process is stimulated in a way that sets the design process in motion. Old ideas can be recycled in new ways, classic designs can be reworked, and styles that are particularly admired can be adapted and refined. Inspiration can be drawn from a library of materials, or result from discussions. More than one winning design was conceived in a pub, where no mental boundaries are imposed by the studio walls, and discussion can range freely from the mundane to the exotic. Of course, in-studio discussions can also yield good and even extraordinary results.

Most designers have an insatiable curiosity about all aspects of design, and will pick up and examine packs at every opportunity. Shopping trips turn into small store surveys, and holidays into pack-collection exercises. Exhibitions are another opportunity to see design at first hand. These may be design exhibitions, graphic-arts exhibitions, gift fairs, hardware exhibitions, or exhibitions related to any topic in which the design group is currently interested.

The fine arts are also a rich vein of inspiration for designers, and some artists' work is used in adapted form in packaging design. The L'Oreal Studio ranges of toiletries and hair-care products have a strong link with the work of Mondrian, an artist whose work has also been used as a basis for corporate-identity work. A knowledge of art, art techniques, and, more importantly, the moods it can create gives the designer a background from which design ideas can emerge. This is also true of photography, and the work of photographers can often have a direct relevance to projects. Photography is a superb medium for creating mood, and, in many pack designs, it is the capturing of a mood that becomes the essence of the design. Fashion in all its forms can also be an important design ingredient. It can be a mistake, however, to produce designs which are too closely aligned with current fashions, as fashion is transient and dates quickly. Nevertheless, if the brief calls for modernity, it may be that elements of currently fashionable designs can be incorporated.

Designers also look backwards to design work carried out in times very different from our own. There are, of course, fashions that are retrospective. The 1960s concept of fashion, for example, may be mirrored as being representative of easier times, rebellious youth, flower power, or whatever other images the 1960s can evoke. A 1960s Cadillac included in a current graphics treatment can evoke nostalgia, but also a feeling about a time when people had more fun. People who lived through the 1960s will know that the current concept of that time is distorted. People look through rose-tinted glasses at the good times of the past, and not at its many drawbacks. This is the nature of nostalgia. In this example, it is the contemporary perception of the 1960s that is important, and designers and marketing people know that the 1960s has a particular appeal for today's youth that can be successfully harnessed to convey a defined imagery. There are many other eras from which the designer can seek inspiration. Even though there are now few consumers who were young in the 1920s, art deco and the slim elegance of 'flappers' can be adapted to today's design requirements to conjure up a concept of elegance, and sophistication spiced with a dash of the decadent.

The inspiration for design thus comes from a variety of sources, past and present. Originality is still required, however, to identify suitable design sources, and then translate the concepts into an acceptable working solution to a design problem.

BRAND STRENGTHS, STANDOUT AND IMAGERY

A brand can be one of a company's most important assets. When a customer base is built on brand loyalty, this is indeed the case. In all packaging projects in which the brand is one of the key factors, the brief should make the importance of the brand clear. Surprisingly, some marketing groups become so familiar with the brand that they start to forget its value. Established brands which are successful in a market cannot afford to stand still. As with most aspects of business, products either move ahead or slip back, and standing still is seldom an option. Slipping back is not an option either if one is a marketing executive who wishes to keep one's car, job and pension. Brands have to progress, and there must always be moves afoot to ensure that they do. One of the most cost-effective methods of promoting a brand and achieving greater sales penetration is to address its packaging. The pack may be for an existing or a new product, but, in either case, the brand identity must be maintained. Shoppers who are looking for their favourite brand of cornflakes do not wish to spend time searching the shelves for it. It has to be immediately obvious and recognizable at a glance, and it must reassure the customers that it is their old, well tried and approved brand. Deviating too far from what the consumers expect will arouse suspicion. Consumers may feel that someone has been tampering with their favourite product, and start looking for alternatives. Brands are vital to product success, and designers have a duty to maintain their strengths, and use them to reinforce brand recognition.

In one example, a design consultancy was employed by a major automotive manufacturer. The brief embraced packaging, but also extended into retail-showroom design and other areas. The consultancy conducted a protracted and expensive study of the client's current design situation, and then made recommendations. One area that it felt should be reviewed was that of the company's logo. There were good design reasons for this, but, no matter how important these were, they were completely overshadowed by the need to maintain the brand identity. The logo had been devised by the company's originator, and had appeared on the bonnet of every car produced by the company since then. There was a horrified silence when the design team challenged the logo, and this was followed shortly afterwards by the design team leaving the presentation room for the last time. Designers must realize that there are historical reasons for brands developing in a certain way, and that many graphics devices that are intimately associated with the brand are sacrosanct, and not up for negotiation. It is quite usual for a design team to be presented with a brand which has to incorporate certain elements which *are* the brand. The element is most often a logo, but there can be other important features. Some brands rely on colour. An extreme example is that of cigarettes. In the UK, the restrictions on advertising have led in some cases to the development of colour as a cigarette brand, without there being any further brand identification. The advertisements have, in this case, become progressively more abstract in character, and yet brand recognition by consumers and nonconsumers remains high. For example, Silk Cut has used purple as its brand signature in cigarette advertising. Its advertisements depict a range of purple objects in association with cuts, for example a purple silk sash being cut, or a reel of purple cotton and a pair of scissors. The colour takes the lead in the advertising, but in the packaging it is restricted to a purple square on an off-white pack. The colour link between the pack and the advertising is apparently strong enough for the brand identity to be maintained. The chocolate manufacturer Cadbury has used a purple foil wrap on its products for many years, and purple has been adopted as a house colour. Cadbury also exploits the colour (which is similar to that used by Silk Cut) as a brand-identity device in advertising. It also uses a similar advertising technique, although there are no restrictions on the advertising of chocolate. Brand is often associated with colour, and, more overtly, with logos.

Using colour as a brand identifier poses its own problems when the brand becomes an umbrella for a variety of products.

For many years, Oxo, which is a well established brand with a powerful logo, was associated with a red pack. When the company wished to diversify from its traditional beef-stock cubes into other varieties (vegetable and chicken), there was a problem in trying to differentiate the varieties while maintaining the brand identity. Colour was the natural choice to indicate the flavour differences, but this could have diminished the brand identity. Fortunately, the Oxo logo was strong enough to survive the change. It was even slightly 'doctored' to increase the standout, and so the colour change was less of a problem. Many market sectors are defined by colour, and, to introduce new products into them, designers almost certainly have to follow the colour code originated by the brand leaders. Low-calorie products (slimmers' products) are, for example, nearly always depicted with a white background. Chicken-flavoured products use yellow, vegetables green, and so on. If the brand identity of the designer's product coincides with this code, and the product is in the same market area, it is possible that there may be some confusion. Differentiation by colour becomes difficult, and the designer may wish that colour had not become such an important property of the brand.

Branding is most often associated with logos. It is the overall shape of the logo that is important; most consumers cannot define the detail. The consumer is only really interested in identifying the brand, and is not particularly interested in the finer points of design. The shopper wants to locate his or her favourite brand quickly, and move on. It is only designers who become fascinated by the finer points of logo design. When a logo starts to show its age in design terms, this is not usually just because it looks dated, but because it starts to lag behind its competitors in terms of standout. A typical logo-design assignment would entail the modernization of a logo, to make it easier on the eye, but primarily to make it more visible. For brands that have been around for many years, this would probably require a subtle touch. Consumers must not, at all costs, be alienated by new design. They should not notice that change has taken place at all, although they must subconsciously react more quickly to the new logo than the old logo. The Shell company's logo has evolved constantly since its conception, each stage of the evolution being a small one which would not be noticed by the average consumer. This is the trick that designers have to bring off; increasing the standout and making the logo more acceptable by modern standards without anyone noticing the change. The Coca-Cola company had an even more difficult task when the name of its product, which had been an established brand leader for many years, was abbreviated by the public to Coke. It also faced changes in the pack format when the competition from canned drinks forced a move towards cans, and away from the famous waisted bottle. The existing style of the logo was retained, but the wording was substantially modified to 'Coke'.

The establishment of brand identity through the use of shape is yet another option. Heinz adopted the tombstone shape on its labels many years ago, and it has managed to adapt the shape to fit an expanded range of products. It would be an unlikely choice of shape if the labels were designed today. It is a device that has been inherited from earlier times, and yet Heinz feels, probably correctly, that it identifies its brand almost without the need for further text. This is the essence of achieving brand recognition: producing a unique device, whether through colour, logo or shape.

Established brands such as those discussed above need careful treatment if the brand identity is to be maintained. Updating and improvements to the standout have to be carried out without destroying the brand values. The creation of a new brand is a quite different matter. The designers are usually presented with a brand name, although the studio will very often also be required to consider suitable names. Another approach to naming a brand is to employ a specialist company that originates names and that subsequently carries out a search to ensure that the name has not been registered by another company.

A brand name is just the start of a branding exercise. The vital ingredient is imagery. The client must provide as much background information as possible to allow the designers to build up a mental picture of the brand qualities, the consumer and the market. Competing products must also be considered, and relevant trends identified. The brand may need to be close to those of the competition, or it may be necessary for it to be distanced from them, depending on how the client's marketing group see the situation. Often, there is more than one brand name, and the design team proceed with several candidates, experimenting to see how they work within a design format. Long names are a problem, because the lettering may be too small on tall narrow packs. Particular letters may also cause problems, depending on where they appear in the brand name.

Consider a hypothetical brand name, Uma. The type of treatment that a designer will consider for the Uma logo if the product is a liquid detergent is quite different from that if the product is a skin-care lotion. The imageries required for these two different product areas are obviously not the same, and the existing brands in these areas will influence the design. The designer might use a fairly bold treatment for the detergent, but a much softer, more delicate approach for the lotion. Both will have to stand out, but to different degrees. Uma the detergent might have to shout, but Uma the lotion will have to seduce. The distinction between the two will be created by the choice of typestyle, colour, possibly texture, and certainly background colour. The situation is further complicated if a range of products is planned. It is better to establish whether this is the case at the start of a project than to attempt to make adjustments later on. If the detergent brand name is going to appear on associated products, it may work. If the product range is too diverse, it may not. The Uma logo may be acceptable on a household bleach, but it may not be compatible with the Uma logo on a household soap.

Brand development typically starts with experiments in sketch form, in which the designer explores the possibilities. In the liquid-detergent example, it is likely that the designer will acquire samples of all the competing products. He or she will probably also examine detergent advertisements in magazines published in other countries.

Reference material is used to provide an insight into the product area, and to identify the imagery that is currently accepted as the norm within the liquid-detergent market sector. Its use does not mean that the designer will slavishly copy what has been done before, but he or she will identify existing designs that work better than others, and analyse the reasons for this, albeit in an informal way.

The designer coming to grips with the product and its imagery will be a key factor in the project's success. In studies of this kind, it is not uncommon for 100 potential solutions to be considered. Even with a word such as Uma, there will be typefaces, colour and backgrounds to consider, and possibly some holding device to prevent the brand name from floating on the pack. Letter styles and proportions, italic, upper-class and lower-case type, drop shadows, vignettes, keylines and other graphics devices will all be explored. It is most likely that the letters will be hand-drawn, and not taken from any existing font, to create a personality for the brand that is unique and unambiguous.

To make each design variant more relevant and easier to assess, a flat cutout of the bottle may be used in an appropriate colour, with the branding options being mounted on clear acetate overlays. This also allows positioning experiments to be carried out, perhaps with the logo being run at an angle. The designs that look promising at this point are often transferred to computer; they are scanned in from roughs, and then 'hand-drawn' on screen. This allows further manipulation, again with a simulated bottle being used as the background. There are many options, and it is not hard to understand why there could be 100 or so variants during the early stages of a project.

The branding will not be seen in isolation, and there will be other design elements to consider. How the proposed brand will work in juxtaposition with these will also require exploration. Detergent designs are notorious for trying to squeeze a great deal of information onto each panel, which often results in a cluttered pack, with text and branding fighting each other. The relationship and importance of the text to the brand has to be established. The client will have stated the ground rules for this, and may insist on retaining too much text, in the designer's opinion, on the front panel. The text may need to state that the product is 'nonbiological' or 'new', or give a contents size, or transmit a sales message, or state that the product is for 'automatic' machines, and so on.

There may be other areas of branding to consider, such as a brand device. On Fairy Ultra, which is a liquid detergent produced by Procter & Gamble, a chubby baby in a nappy strides across the pack. The same child appears on other Fairy branded products, although it is used in slightly different ways. In the detergent example, the designer may wish to adopt some visual device which will provide brand identification. Devices of this type have the advantage that they can be easily 'read' by consumers, and, if the product is successful, they become the link between the product variants. Visual recognition of a symbol or device is more rapid than the reading of text, and so the device becomes a valuable brand property. The simpler its design is, the better. Complicated images cannot be scanned quickly, and something that can be recognized immediately is required. The studio will explore many methods of incorporating graphics branding onto the pack, bearing in mind the future plans for the brand. It could be of little use to develop a device that was so specific that it could not be transferred from the product to other product areas, if this was the future marketing plan.

The need for the involvement of many disciplines in packaging design has been stressed throughout this book. When a new brand is developed, or an existing one is revised, an input from the advertising-agency team is of particular value. They will often assess the product's packaging to see how the pack could be featured in advertising campaigns. Brand devices originated in packaging work have sometimes been a source of advertising inspiration. For example, Cadbury's Marvel is a powdered milk that is vitamin-enriched and therefore of dietary interest. As the product is low-fat, and contains vitamins, the packaging designers incorporated the silhouette of a female figure on the pack. This was picked up by the advertising agency, and animated on television advertisements to reinforce the healthy aspects of the product, and to reinforce the branding device.

COLOURS

The use of colour is very important in graphic design. When selecting colours, the designer will consider standout, to ensure that the pack will be visible on the shelf. There are some constraints to work within, however, and the brightest, boldest colour is not always the most suitable for the design.

Dark, rich colours are associated with quality. For luxury goods, for example a box of liqueur chocolates, deep reds or purples can provide the imagery required. Metallic colours, for example golds and coppers, may also be suitable. However, the danger with metallized finishes is that, in some lighting conditions, they can look black, and become recessive. Black is also associated with quality (often a rather decadent quality). Its association with death in the West limits its use, and yet Western consumers accept it on Eastern products, particularly Chinese or Japanese foods, or an interpretation of them. In the Middle East, white is associated with death. Market understanding is essential when colours are selected. Green was seldom, until recently, used on packaging other than that for vegetable products. The link between meat products and green was an unhappy one, carrying connotations of decay. Now that environmental groups have green as a flagship colour, there is a change in attitude, and the colour now sits happily on meat-product packaging.

Colour is an emotive medium, and has latent meaning for everyone. Designers, who have an understanding of colour and its use, are able to advise on the most suitable colour combinations for product packaging. The issue is one of imagery, and choosing appropriate colours to convey the required imagery. Often, it is necessary to use particular house colours. There may be unwritten colour rules to follow for particular market sectors. The design team should be allowed some latitude, and then the correct colour for the product will be chosen. Potential designs fail for all sorts of reasons, but colour seems to be the least significant of these.

METHODS OF WORKING

In the development of branding and brand devices, the background against which these are tried may be complex. It may not have just one colour. A vignette may be used in which one colour or tone merges with another. This is traditionally accomplished by the use of an airbrush, which creates striking effects on a visual. The problem is that achieving the same effect when the vignette is translated into print can be difficult. In fact, it is never known whether a vignette will be successful until print trials have been conducted. This is not to say that vignettes should be avoided, but that care should be taken about their use in the initial stages of a project. Before anyone becomes too enthusiastic about a vignette background as a result of seeing an airbrushed visual, questions must be asked about its viability in print, or a print trial should be organized.

Most design teams start projects with pencil or pen sketches. Depending on the ability of the designer, these may be good enough to convey conceptual thinking to the client. Mostly, they will not be adequate for this task, and some visuals will be needed. Some design companies employ visualizers whose task it is to take selected designs, and work them up into visuals that are of a sufficiently high standard to allow the studio to communicate with the client, and to allow valid judgments to be made. Such visuals are often produced using marker pens, and, properly executed, are excellent. They can almost be small works of art, having the freedom of movement and flair that is associated with some schools of art. It is not uncommon for clients to display such visuals proudly in their offices. With the advent of computer systems that are capable of doing the same task, marker visuals are often replaced by computer-generated drawings. These can be displayed on screen, or as colour printouts, or, with some systems, as slides or photographs. The computer is a demanding tool, and it will give extremely accurate results, provided that the correct information is fed into it. The problem is that, in the early conceptual stages of a project, it is more important to develop a range of ideas than a few highly finished ones. It is possible to produce visuals several times faster using markers than it is using a computer.

When the project progresses further, and highly finished mockups are needed, the computer becomes a much more useful tool. Even then, if airbrushed backgrounds are part of the design, it may be quicker and better to produce these conventionally, and to apply text and logos generated by computer to them in the form of rub-down transfers. There is little doubt that the electronic generation of designs will become the norm, with new and faster systems being developed which will also be more compatible with the way in which designers work. There is also an advantage to the use of computers in that whatever they produce is stored, and can be recalled to form the basis of artwork later in the project. This is not an advantage at the conceptual stages, when many designs, most of which will not be used, are generated.

There are two areas of concern for any client. First, the client pays mainly for creativity, and visuals should be a communications medium, not an end in themselves. The client should not have to pay for excessive time spent in preparing visuals, when that time could be spent on cracking the design problem. If the visuals are too slick, it could indicate an incorrect emphasis and allocation of time and resources. Second, all visuals should be prepared to the same standard, otherwise comparison between design candidates becomes difficult. It is easy to be swayed by a visual that is more highly finished than its companions, simply because it has a more professional look. High finish is not the same as good design.

In the early stages of a project, some design routes may involve photography or illustrations. Instead of spending time and money at this point, studios may use existing photographs cut from magazines or taken from any other source. This is simply to convey the essence of the concept. Illustrations may be handled in the same way, but they are more often properly executed in-house by one of the design team. They will be quickly prepared in any suitable medium, perhaps crayon or water colours. The work is bound to be 'loose' because of the speed of production. On occasion, this is exactly what the client likes. Some initial illustrations have been preferred to later fully worked examples. Should design candidates including either photography or illustrations proceed to the final stages, it is usual for the final work to be put out to specialists under the art direction of the studio.

The final preproduction stage is that of proofing. This is usually the stage when printer's proofs are examined before the print run is given final clearance. There can be an interim stage, however, when the studio produces its own 'proofs'. These are highly finished mockups that are produced by a variety of techniques, but usually by computer, from which they are output as dye-sublimation, Chromalin or laser prints. The quality of these is improving all the time, and, for most purposes, they are accurate enough. Often, the colours are not quite 100% accurate, and some effects may be difficult to achieve. They do not replace printer's proofs, but they can be suitable for design approvals or for market-research material.

Graphic design is a visual skill about which much has been written. To gain a real appreciation of the way in which it is carried out, the methods used, the creative ability of the designers, and the technical expertise involved, it is necessary to see it in progress. It is recommended that, whenever the opportunity presents itself, potential clients should visit a studio, and witness design in action.

9
Artwork and reproduction

When packaging projects reach the stage of artwork preparation, there is all too frequently a last-minute panic to complete the project on time. The artwork stage occurs fairly far down the project path, and, by then, it is common to find deadlines approaching rapidly. The pressure is on to make up for earlier delays in the project. In these circumstances, it is not surprising if errors creep in, checking and proofreading are skimped, and the results are less than satisfactory. If a proper project plan is drawn up, this problem need not arise, but experience shows that, whatever the plan shows, it is artwork preparation that is often done under the greatest pressure of time.

The days are largely gone when artwork departments worked on drawing boards, using transfer lettering and producing pasteups. Now, artwork is almost always electronically generated, with computer screens replacing the drawing boards, computer fonts replacing the shelves of transfer lettering, and discs replacing the master artwork. Earlier skills have been replaced by new, equally demanding ones. Artwork production remains as much a specialist skill as it has always been, despite the high-technology hardware and software that is now available.

The purpose of the following text is not to detail the physical process of artwork and reproduction, but to assist those involved with packaging-design projects to understand and overcome the problems that most often occur.

COPY PREPARATION

The primary requirement when copy is written is to make sure that it is accurate. All the copy supplied to the studio that is responsible for producing the artwork should be checked, approved, and legally cleared before it is released. Although it may be a simple matter to make corrections later in theory, in practice, this can result in considerable additional costs and delays if the studio has to reset the copy, particularly if it creates additional lines. Copy presentation should also be accurate. The copy should be typed, and good-quality copies supplied to the studio. Poor-quality faxed versions scrawled over with handwriting are too common. This is not good practice, and easily leads to misunderstandings.

The artwork studio cannot be held responsible for checking copy. Their job is to set what they are presented with, although, of course, if they do spot errors, they will draw this to the attention of the copy originator. Copy may be in languages that are unfamiliar to the artworker. This is not usually a problem, as the copy will be set character by character. Non-European languages (Arabic, Chinese and Japanese for example) may, however, need the specialist attention of a separate outside agency.

There is no point in numerous lines of copy being written if the space on the label that is available for copy restricts the text to one or two lines. There is a tendency to try and force as much text into the available area as possible, resulting in a crowded mass of small-sized text. The object of including any text is for it to be read. If an attempt is made to extol the product's virtues by the inclusion of too much information, the consumer will simply not bother to read it. Text must be kept to a minimum on packaging. This makes the task of reading it less daunting. It is also then possible to increase the type size to a size that is legible, even for customers with less than 20/20 vision. With environmental pressures to eliminate secondary packaging, and legal obligations to include product data on many categories of product, there are now greater demands on the available space than there used to be. Nevertheless, cluttering a pack with unnecessary copy is a mistake. Designers will fight for, at the very least, a front panel with carefully considered and balanced graphics that includes enough space for the graphics to 'breathe'. Communication is thus immediate, and does not rely on close-up inspection. When this has been achieved, and client approval has been obtained, it is particularly disappointing if the concept is ruined by spurious copy plastered across the space. The message is simple: copy must be kept to a minimum.

When the copy is being prepared, the copywriter should have details of the panel sizes. Areas for bar codes and other data that must be included should be identified. This information is best presented in a drawing format, and it should be actual-size. The studio will be pleased to provide this information. It is then not difficult to visualize the overall impact of the proposed copy. The artwork studio can also provide an indication of the finished appearance at the recommended type size by setting a panel in 'Greek' (i.e. Greek characters are set to indicate the final appearance of the text when no copy is available). The effects of increasing or decreasing the text size can similarly be seen at minimal cost. This setting provides a useful guide as to the number of words that can be included in a line, and the number of words that can be accommodated in each panel, for different type sizes. In more critical circumstances, for example when the text is an integral part of the design, more detailed guidance from the design studio will be needed. For example, a front panel that has an area that has been carefully arranged for the product descriptor is really part of the design. Designers have sometimes worked with one language in this area only to find that another version is later required that uses a language which occupies more space, because of increased lengths of words in that language. This problem should be avoided by the provision of all the required language variants early on in the project. The designers and artworkers can then liaise to ensure that the demands of the graphic design and the copy requirements are not in conflict, which is not always an easy task.

It is important to ensure that the copy is written, approved and submitted to the artwork studio at the scheduled time.

There will be a cutoff point after which failure to supply copy will mean a delay in obtaining the printed material. The copy deadline, however, should be as far in advance of this point as is possible. Far too often, the copy preparation is left to the last minute, and mistakes then creep in, often entailing expensive alterations to films. There is no excuse for this, but it happens in too many design studies.

When good clean copy has been received, the usual artworking practice is to set the copy, check it, and clear the overall appearance with the designers. It is a design responsibility to select the typestyle, this having been a part of the graphic-design study that has been approved by the client. When the designers are happy with the result, a photocopy of the set panel is usually sent to the client for approval and final signing off. There should not be any need for text alterations at this point, as the copy should already have been agreed. Surprisingly, however, there are often copy changes, as opposed to corrections, at this point. There is then a repeat of the copy cycle, and the expenditure of yet more time and money.

ARTWORK PREPARATION

Artwork is almost always produced electronically. A number of systems are available, but Apple Macintosh systems dominate this market.

The process of preparing artwork on an Apple Macintosh type of system is shown in Figure 9.1. Once the graphic design has been approved, the artwork department can begin its contribution to the project. The first step is to consider how the design is to be printed. The artwork department analyses the design in conjunction with the design team in terms of the following:

- the use of photography and/or illustrations,
- the number of colours,
- the print method to be used.

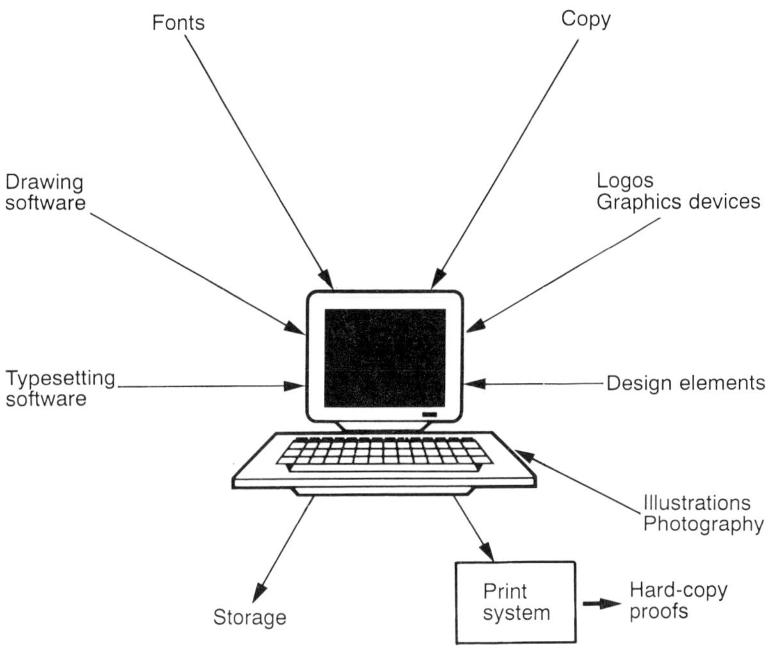

Figure 9.1 Artwork system of Apple Macintosh type

Panel sizes that give accurate dimensions of the print area available are usually provided by the 3D designer if the pack is a new one. If it is not, sizes can be obtained from suppliers, or the existing packs can be measured. The design team will also provide sketches that show the positioning of the design elements, including logos, photographs or illustrations, and text.

Often, it is possible to start work on logos and other graphics devices before the copy is finalized. Existing logos can be scanned in from a corporate-identity manual or existing printed material, and then graphically cleaned up using one of the many software packages designed for this task. Logos and any other constant design elements can then be positioned on the panels, and correctly located. When four colours are used, each colour is assigned to a 'separation' that corresponds to one of the four colour plates that will be used to print the job.

The design team will have selected a typeface which they feel is right for the desired overall pack presentation. There are vast numbers of typefaces, and they are nearly all available on disc. However, the artwork studio may have to buy in special typefaces, which may take a few days, and will incur additional cost. Using suitable software, the type can be set to the required size. Artwork expertise is then applied to the line spacing, word breaks etc. The artworker will know what print process is going to be used, and select an appropriate type configuration and size to avoid, for example, the filling in of reversed-out text. It is here that conflict between the client and the artwork department can arise. The artworker knows what can be achieved, while the client may press for the inclusion of additional copy which may not reproduce well when it is printed. The designers may also pressurize the artwork department by insisting that the design concepts are not compromised. A satisfactory middle ground will eventually be attained after some discussion.

Photography and illustrations are often only indicated on the artwork in the form of areas where the repro house will strip in the colour-separated artwork. This is a specialized process requiring complex scanning equipment, and it is usually outside the remit of artwork departments. Colour transparencies are usually supplied to the repro house with the artwork, and the repro house then completes the process.

Colour tints and vignettes are produced by the artwork studio. In vignettes, the tone changes gradually from light to dark, or the colour changes gradually from one to another. Vignettes are used in many types of packaging design, and they can look very effective. For the artwork studio, however, they represent a problem. The vignette has to be interpreted, probably from an airbrushed original design. This is not an exact science, and the translation of the designer's intentions relies on the experience of the artworkers. There is no guarantee that the result that the artwork team considers to be correct will have the finish that was envisaged by the designer, and that has been communicated to the client. The only effective way of ensuring the exact matching of vignettes is to conduct print trials. This is a costly exercise, as plates (for litho printing) or cylinders (for gravure printing) will have to be made up for the trials. If this option is not available because of budgetary or time constraints, the appearance of the printed vignette will be a guess. Marketing departments, in particular, often find it hard to understand that there can be so much doubt about the final result. However, even proofing systems are unlikely to yield results that are 100% the same as the printed version. Unless the budget allows for print trials, it is necessary to trust the expertise of the artwork department, or avoid the use of vignettes.

Before the process moves on from the artwork stage to film making or other forms of origination, it is usual for proofs to be produced. Usually, this is a two-stage process. All the print detail is stored in colour-separated form on disc, and the first stage is to obtain the client's approval of a black and white proof. It is not difficult to produce laser copies of artwork. The problem lies in their interpretation. It is easy to produce a black and white composite print in which each colour layer is superimposed in the correct order, as would be the case in a finished printed example. While this gives a good overall impression of the design that will be seen, it does not indicate which colours are used within each area.

Laser copies of each layer or colour proofs for each colour are usually also submitted for approval. This enables the colours used for each area to be checked. If these proofs are copies of the actual artwork, however, there may be some degree of distortion. The artwork department may have introduced 'chokes' and 'spreads', which are extended areas of colour that allow for a margin of misregister, which is an inevitable but predictable shortcoming of the printing process. This can cause proof-reading problems. Where text is reversed out of solid areas, laser copies may be accurate enough, but faxed versions may be too distorted to read. Ideally, proofs of the pack should include a composite black and white proof and colour proofs for each colour that do not show the chokes and spreads, all in the form of original laser printouts. Sometimes, and particularly when approval has to be obtained from more than one source, this entails a considerable amount of additional work in the preparation of a special 'nonspread' version for approval that consists of five sheets for each page. Enlarged versions may also be required for fax transmission. Thus the electronic artwork-generation system has to generate large amounts of printout, possibly from three separate files for each pack design. If there are three sets of proofs for each design, an artwork version, an approval version and a fax version, the chance of errors creeping in increases. It is not uncommon for one of these versions to be corrected while the others still include an error, possibly leading to printing errors at subsequent stages.

Whenever possible, there should be one master set of proofs. These should be circulated to all the relevant parties who must check, correct or sign off the proofs of the artwork.

There is a valid argument that colour proofs should be seen. One might assume that they would accurately show what the final pack will look like. Unfortunately, this is not always the case. The accuracy of computer-generated colour proofs is often limited by the process of their production. Dye-sublimation prints, for example, are inexpensive and quick to produce, but the colour reproduction and quality detail does not match the final printed version. Although technology continues to advance, a system that can produce proofs that perfectly match the final printed pack is not yet on the market. Current systems, despite their drawbacks, do have an important part to play. They can provide a close approximation to the final results that may be good enough for proofing and for using as the basis of mockup packs. This is particularly useful for market-research mockups, as small quantities of proofs can be produced at relatively little cost and at short notice.

Printer's proofs remain the best option for the accurate checking of the whole design. A short run of printed material is produced for checking. The correct grades of materials are used, as is the print process that will be used for full-scale production. This can be an expensive step, as plates have to be made, and enough copies run off to ensure correct machine-setup and running conditions. While one may only want to see a few examples for checking, 1000 may be produced before the machine has settled down. Alternatively, a proofing press may be used. Less material will then be used, although the results will not be quite the same as those from the machine that will be used for the main print run. If there are mistakes at this proofing stage, new plates (for litho printing) will have to be made. For gravure printing, the origination of new cylinders will cost hundreds of pounds.

In projects with tight budgets and timescales, it is quite usual to proof 'on press'. The artworker, print technician and/or client attend the first print run of the job, and approve it, or inform the printer of corrections, when it is still on the press. Such proofing sessions sometimes take place in the evening or in the small hours of the morning. Dedication and a strong will to succeed are tested to the full.

The printer must be given the artwork in the correct format. If conventional litho printing is to be used, and the printer is not going to originate the film, the artwork studio must find out from the printer what type of film is required. The following information will be needed:

- Positive or negative working plates?
- Emulsion side up or down?
- Right-reading or wrong-reading?
- Colour print order?
- Bar code supplied, or only EAN number?

The film origination, which is usually carried out by an outside specialist company, can then be commissioned to meet the specification required. Film is often originated directly from a disc supplied by the artwork studio. Increasingly, the data is sent via a modem link directly to the printer, who is then responsible for the origination of the film, or to a specialist film-origination company. The advantages of speed are obvious, particularly when the film originator is overseas. Artwork studios like, if possible, to examine the film and carry out a last check. This is becoming less common with the use of electronic data transmission, and it is now more usual that the stage at which the final artwork is stored on disc is the last point at which checks can be made before the print run.

PRINT METHODS

When it produces artwork, the artwork studio responds to the method of printing; it does not dictate it. Its task is to supply artwork in a form that the printer can use with a minimum of further processing. With conventional print techniques, there are interim stages between the production of the final artwork and the print run. These are film origination, plate making, and makeready. By far the commonest printing process is offset-litho printing, although developments in flexography mean that the use of this process for packaging work has increased.

The most significant changes in printing will stem from developments in digital art and reproduction (DAR). This is the process of producing artwork digitally (this is almost always the case already), feeding this data into a plate-making machine, and hence producing plates for press. There are also developments under way which will eliminate the need for plates. In this system, the press incorporates cylinders with surfaces made up of cells that can be either ink receptors or ink repellants. The cylinders are online to a computer. The image on the cylinder is displayed on a computer screen, and the status of each cell can be changed by the computer in accordance with the digital artwork data, thus producing the desired print results. Full-colour printing can be achieved by using a cylinder for each colour, or even one cylinder with a revolution for each colour.

The benefits of this system are immense, in that the digital artwork can be colour-separated, and then transmitted directly to the printer, eliminating the interim stages. The image on the cylinders can also be changed during the print run.

ARTWORK LIBRARY SYSTEMS

Traditionally, flat artwork has been stored for a time after its first use, by either the printer or the artwork studio. This artwork is needed if new plates are required or minor amendments are to be made, such as promotional or price flashes.

The advent of electronic production of artwork has changed this. Artwork is now retained in digital storage, ready for retrieval whenever required. It is then relatively simple to pull up a piece of artwork from the archive to the screen, as shown in Figure 9.2, and amend it as required, saving time and money. Another advantage of this way of working is that there is less chance of errors creeping in than there would be if all of the artwork had to be reoriginated. For example, there are many cases in which a common format of label is used for a variety of products. Constant details, such as the panel size, logo position and content declaration, remain the same, while perhaps only the product descriptor changes. Library systems thus minimize work while maintaining accuracy.

Some client companies have also installed computer systems that are compatible with those of their design consultancies. This allows data to be transmitted between the two parties through a modem link. Proofs of artwork can be sent to the client, who can examine them on screen, and the appropriate action can be initiated. This can extend to simulated 3D versions, for example that show a 'label' wrapped around a 'bottle', or show shelf-display comparisons. Similarly, the client can transmit copy via modem directly to the studio, again with benefits of time saving and accuracy.

Printers who store artwork may have a room full of traditional flat artwork, which has to be kept in case new plates are

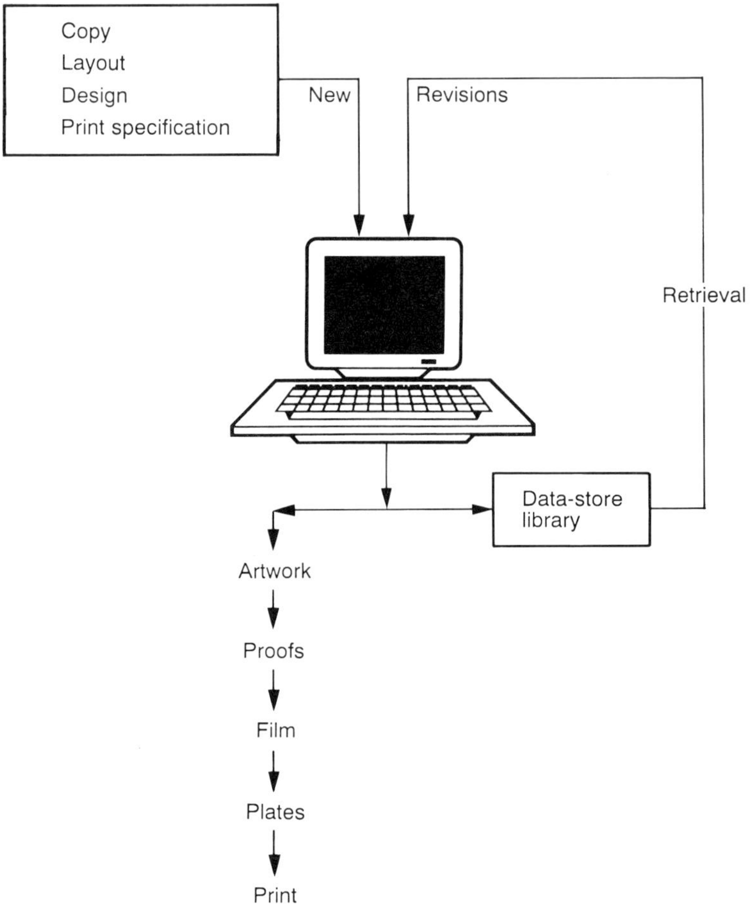

Figure 9.2 Artwork library system

required. It is often the case that pieces of new artwork for updated versions of a product are stored with the old artwork, because there is a reluctance to discard the old artwork, or there is no client instruction to do so. This not only leads to a proliferation of artwork, but also, potentially, to mixups when producing printed material. Digital systems eliminate storage problems, and offer facilities for cataloguing stored material, and ease of retrieval.

10
Strategic packaging design

MARKET RESEARCH

A few issues relating to the organization of a packaging-design study remain to be considered. It has been suggested above that market research should be carried out at the conclusion of the second stage of a study. All the conceptual work has been carried out during the first stage of the project, the recommendations have been made, and those candidates worthy of progression have been refined during the second stage. However, there could still be two or three options that are liked for various reasons. Often, when there is a panel of marketing staff from the client company, there is simply no agreement between them as to which design should be proceeded with.

The purpose of market research is to provide some form of measurement of consumer reaction to the new or revised packaging, and thereby to guide marketing staff on the choice of pack. There are two schools of thought about market research: some favour it, and some do not. Its supporters regard it as a means of obtaining consumer input to the study. It is the consumers who will exercise their judgment and purchase the product and pack if they approve of it, or leave it on the shelf if they do not. Consumers cast votes by their purchasing decisions. Opponents of market research view it as a crutch to support marketing decisions, and a means of delegating a marketing responsibility to the consumer. Both schools can support their arguments by the use of case studies.

Whatever the reader's opinion on this may be, it is obvious that, by the time research is conducted, the design process will be fairly well down the road, and established design routes, for which there will be a logical rationale that satisfies both the design and the marketing departments, will already have been completed. A design commitment will have been made. The problem occurs when the same logic is applied to more than one design candidate, resulting in two or more options that are equally valid in terms of meeting the brief, but different, perhaps, in character. Even if the studio recommends just one design route, the client may not be sure about it, and may also see the validity of other directions.

Design studios are often reluctant to make specific recommendations. All too often, studios are guilty of leaving their options open right up to the last minute, and not being firm enough in making recommendations. A studio naturally wishes to please its client, and it may, if there are multiple design solutions all of which meet the brief, wait until the final presentation before firming up on one or another of the candidates, doing this in response to the client's reaction. This is bad practice, and not conducive to good design. A studio is employed for its professional expertise, and it should have the conviction to support the design it believes in passionately. It should not attempt to back more than one horse during the race, even if it sees by the client's body language that the odds are changing. A studio that keeps its options open in this way may be enthusiastic about market research because of its inability to make decisions.

Marketing staff are also professionals, and, by the time firm design directions have been established, they should have exercised their judgment and influenced the design process accordingly. A degree of personal preference is involved when a winning design is selected. No matter how logical a design conclusion may be, there will always be those who prefer a different design. In a situation in which several design candidates are recommended, and there is more than one member of the marketing team, opinions may vary, giving an inconclusive result. This is when market research is often used to assist the team in making the final decision. In other instances, even when everyone agrees on the winning solution to a design study, market research is undertaken as a final check to make sure that the combined efforts of the design team and the marketing team are on target. If the market research reveals a poor reaction, the team should be sacked. In other words, if the team is so sure of its design direction, why check it?

Those who do not acknowledge the benefits of market research must be prepared to stand by their own decisions, accepting the glory if the design is successful, and taking the consequences if it fails. Detractors of the market-research concept point out that this is what they are (highly) paid for in the first place.

Market research is a highly specialized area, and most design consultancies acknowledge, as specialists themselves, that skills that are very different from theirs are required if useful research is to be conducted. Few, if any, design groups also carry out research, but most are happy to work with research companies and to recommend suitable ones that are expert within the packaging area. There has to be close cooperation between the studio and the research company over the provision of research material. The studio will also be keen to see the results, and will wish to be present at any market-research debriefing sessions.

The success of market research depends on the relevance of the questions posed to respondents, how the research is conducted, and, for reasons of accuracy, how many respondents there are. The nature of the stimulus material is also important, and this is one of the primary contributions of the design studio. It can often also assist in the formulation of the questionnaires. The timing of the research may be critical, and this may mean that there are restrictions on the type of stimulus material that can be produced within the given timeframe. In studies involving 3D design, it is desirable, if not essential, to prepare models of the pack. Consumer groups may be asked to judge the 'new' pack against the existing pack and/or competing packs. It is essential that comparisons should be 'like-for-like', and so the standard of presentation must be the same for all the material. This can entail the production of models for new, existing, and competing packs, which could amount to a considerable amount of model-making time. Variables have to be eliminated as far as possible from research. The pack structure will probably be considered alone, without graphics or colour being used. All the packs will be of one colour in this situation. If several locations are involved, the number of models increases, unless locations can be used consecutively for the research sessions. The problem with this is that the timescale is extended, and costs can also escalate rapidly, especially if the models have to be handled by respondents. The construction must then be robust enough to survive repeated handling. The models may also have to be made to a higher specification. For example, the weight and centre of gravity may have to simulate those of the actual product. Details of particular relevance to consumers may also have to be incorporated. Devices such as pouring spouts and dispensing closures may be important features of the design, and it may be necessary to fabricate them specially for the market research. This is often a difficult task, and it increases costs.

Photographs or computer-generated images of containers and packs may be cost-effective alternatives. If photographs are used, however, some form of physical 3D model will have to be made. However, it will not need to be of such a high mechanical standard as it would if it were going to be handled. If photographs are used, the model maker will probably 'cheat' by only finishing the panels which are to be featured in the photograph, which will save time and money. Computer-generated models exist in electronic form only, and do not have any physical 3D counterpart. The model is constructed on screen, using the dimensions from a working drawing. At this stage, it is often a wireframe construction. The model is then rendered with whatever finish is required. This is a simplified description of the process. Depending on how sophisticated the results have to be, the process can take several days to complete. The final results can be remarkable, however, allowing the model to be rotated around any axis, and incorporating any lighting, reflections and shadows that may be desired. The computer can light a model in the same way as a photographer can, and the output can look exactly like a photograph. The use of flat material to depict 3D forms is limited, and does not compare to the hands-on trialling that 3D computer modelling can provide.

New methods are also being developed for producing physical 3D models directly from computer-generated data. CAD/CAM techniques are not new. They have been used for some time in production engineering to produce tooling and parts. Similar systems are available for producing models. These often use wax as the model-making material. The models can subsequently form the basis of castings in metal or resin. Stereolithography is an even more sophisticated technique. This uses a laser to polymerize a liquid resin. The laser is controlled by computer, and there is one beam for each of the three 3D planes. The beams are projected onto a transparent tank of resin, and the resin solidifies where the beams meet, forming a solid structure. These accurately produced structures include hollow sections, and they can form complete containers within the bath of resin. This technique was developed for the production of prototypes, and it is currently probably best suited for that purpose. The equipment is not the type of thing that studios are likely to have in-house, but specialist services do exist which studios can use, at a price.

For research into the impact of graphic design, the ideal stimulus material consists of printed mockups. It is often impractical to incur the expense of a full-scale lithoprint run, and screen printing is often used, which is a cheaper and quicker option. For most research purposes, computer-generated proofs or a combination of computer-generated and hand-finished material is sufficient. When multiple sets of stimulus material are needed, 'photographs' are often used, either in the form of conventional photographs, or computer output.

Whatever methods may be used for the production of material, the research should be designed to provide clear results.

Unambiguous results that clearly identify a winning design are required. This is not always the case. One design may be preferred for some of its qualities, and other designs for different reasons, but no one pack emerges clearly as the preferred route. When this happens, the team are virtually locked into the system, and they have to take the comments on board, modify the packs accordingly, and commission a second round of research.

It is an interesting and nail-biting exercise for designers and marketing staff to sit through a research debriefing waiting to hear about the consumers' reaction to their work. It is equally interesting to attend one or more of the research sessions, and listen to respondents being interviewed. They will often talk around the questions, and raise some points that are not covered by the questionnaire. Side issues are raised, for example the stackability of the pack, that are practical points that affect the design, but are of no direct concern to the consumer.

Unlike most aspects of design, research is based on a record of responses to questions, and these responses can be tabulated and measured. They must be interpreted particularly to determine what information is significant, and what suggests changes that are only desirable. Research results do not have to be accepted, but it is a brave marketing person who does not do this. Following the research, some design refinement will inevitably be required. It will hopefully be minimal, and it is most likely to be confined to increasing a type size or some other minor adjustment. Significant differences between research locations may entail further work, and a second stage of research. One stage is usually sufficient, however.

COMMUNICATION

The need for effective communications at all levels and between the departments involved in the study has been stressed. One of the primary means of communicating packaging details is through the medium of the packaging specification. This document is the basis of all packaging work. It details the materials, the style, the dimensions, the typefaces, and any other information that will be important in the production of the packaging. Specifications are primarily a means of communication between the client and the supplier, ensuring that the final pack is exactly what has been designed and approved. The specification is also a means of storing information.

There is little point in just storing information. The information must be accessible and easy to retrieve. A manual system of specification storage is perfectly feasible, but it is obviously better and more suitable to use a computer-based system. The specification number may seem a trivial piece of information, but it can be used as the key field in a specification system, and some thought needs to be given as to how it can best be used. What is needed is a system that can retrieve documents, and sort them by date of issue, materials, style, dimensions and all other parameters of the specification. The advantages of being able to do this are significant when package rationalization is being considered, as should be the case. It should be an ongoing task to eliminate sizes whenever possible and maximize the pack production quantities, as this reduces costs and inventories. Sorting by specification number will also allow obsolete items to be deleted. This is important when specifications tend to proliferate, and soon clog the system if not deleted.

The other significant document which comes under the control of the packaging function is the packaging standard. This is another document that acts as a communication between the packaging function and the production function within a company, and it may have different names in different organizations. The purpose of the standard is to inform the production department about the packaging materials that are required to accomplish a particular packing task. It may define, for example, the number of products per carton, the number of cartons per outer, the number of outers per pallet, and so on. The standard is an active document that must be easy to retrieve so that it can be modified and updated, and so that deletions can be carried out.

Drawings are also communications documents. Each specification may incorporate a sketch or drawing on which details of the main characteristics of the pack are given. If a more complex drawing is required, the specification will probably only refer to it. Design companies do not normally have the expertise to produce full-scale engineering drawings. Often, the pack supplier will originate such drawings, as the supplier knows what information is required for each machine in the form of technicalities, draw angles, tolerances etc. The design studio may need to provide two standards of drawing: a drawing that shows the principal dimensions of the pack, and one that shows any special features. These will probably be sufficient for a carton supplier, say, to work from. The supplier will add the details, and produce accurate drawings. For a more complex item, for example a blow-moulded container, the drawing will have to contain more detail. It should also show the actual dimensions of the pack. This is not as easy as it may seem. The calculation of the volume of a nonregular container is difficult, and there will be manufacturing restrictions of which the design studio may not be fully aware, such as the depth of the base punt that can be used. Therefore, a studio drawing will usually be as accurate as possible, and incorporate as much detail as possible, but it must still be regarded as an interim document, i.e. as a way of communicating with a supplier that works towards the final pack drawing that is produced to the correct engineering standards. Software packages are available for all standards of drawing. Systems that incorporate plotters are difficult for designers to learn, because coordinates are used to define the position of the pen. This is a system that is alien to designers, but familiar to engineers. However, whether the drawings are produced by computer or by hand, their purpose is to act as a useful means of communication.

THE FUTURE

In the future, more emphasis will be placed upon the environmental aspects of packaging. Of all the influences on packaging, this will probably be by far the most significant. EU-driven directives will require all governments to take this issue seriously. In the meantime, so that UK companies can trade in other countries that do have legislation in place, designers in the UK must adapt to overseas methods of environmental control. In many ways, companies have already been forced to preempt legislation, and it is already common to see packs in all European countries with the German Grunnepunkt symbol, even though this may have little significance in the manufacturing company's home country. There will also be a move towards depackaging, i.e. ways of eliminating packaging altogether will be considered. This will be compensated for by changing point-of-sale conditions. Moves in Germany already point in this direction, and there will be further areas where no packaging is used. Far from making designers redundant, this is a fresh challenge to designers' ingenuity, and it is quite likely that the focus of their efforts may move away from the pack to the point of sale, for some product groups at least.

Even with the increase in environmental awareness, there must still be an important role for packaging to play for many categories of goods. The UK is locked into a system of selling mostly via supermarket outlets, where the pack is the only salesperson. This system is likely to expand throughout Europe. Currently, the UK is very much the lead country in this area, with expertise in merchandising and selling that is well in advance of that in other parts of Europe. In countries within the EU, there will be an increase in the sophistication of marketing techniques which will be reflected in an increased awareness of the role that packaging can play. Even in times of recession, retailers and manufacturers alike look for a bigger slice of a smaller cake. The future of packaging looks secure.

Another increasing influence is the expansion of multinational companies, and products being marketed across national boundaries. Television advertising via satellite links already crosses not only national frontiers, but whole continents. The power of television to reach into people's homes is immense. The development of packaging which is adaptable across continents will be the next step forwards, and this will not just be Europackaging, but global packaging. Only a few design consultancies will be equipped to take on this task, and indeed few will be needed. It is quite likely that there will be a shakeout within the design world that will create many losers, and a few rather rich winners.

Successful packaging design is not achieved using a hit-and-miss approach. It requires a strategically planned and executed study that takes into account the market, the views of professionals in many other disciplines, and the constraints of those disciplines. While it is right that designers should provide the creative flair, that flair must be harnessed to practical concerns, and channelled into packs that sell products. Only a strategic packaging-design programme that can take the project from start to finish will yield the right results: the right pack at the right price at the right time.

Index